NO HEAVENLY BODIES

Infrastructures Series

Edited by Paul N. Edwards and Janet Vertesi

A list of books in the series appears at the back of the book.

NO HEAVENLY BODIES

A HISTORY OF SATELLITE COMMUNICATIONS
INFRASTRUCTURE

CHRISTINE E. EVANS AND LARS LUNDGREN

The MIT Press
Cambridge, Massachusetts
London, England

The MIT Press would like to thank the anonymous peer reviewers who provided comments on drafts of this book. The generous work of academic experts is essential for establishing the authority and quality of our publications. We acknowledge with gratitude the contributions of these otherwise uncredited readers.

This book was set in Stone Serif and Stone Sans by Westchester Publishing Services. Printed and bound in the United States of America.

Library of Congress Cataloging-in-Publication Data

Names: Evans, Christine E., author. | Lundgren, Lars, 1973- author.
Title: No heavenly bodies : a history of satellite communications
    infrastructure / Christine E. Evans and Lars Lundgren.
Description: Cambridge, Massachusetts ; London, England : The MIT Press,
    [2023] | Series: Infrastructures series | Includes bibliographical
    references and index.
Identifiers: LCCN 2023007495 (print) | LCCN 2023007496 (ebook) |
    ISBN 9780262546904 (paperback) | ISBN 9780262376822 (epub) |
    ISBN 9780262376815 (pdf)
Subjects: LCSH: Artificial satellites in telecommunication—History—
    20th century.
Classification: LCC TK5104 .E83 2023 (print) | LCC TK5104 (ebook) |
    DDC 629.4609/04—dc23/eng/20230302
LC record available at https://lccn.loc.gov/2023007495
LC ebook record available at https://lccn.loc.gov/2023007496

10   9   8   7   6   5   4   3   2   1

For Frank Evans (1934–2022), an engineer and a stargazer
For Kicki Lundgren (1935–2021), who found beauty in the little things on Earth

CONTENTS

ACKNOWLEDGMENTS

This book would not have been possible without the generosity, creativity, kindness, and flexibility of numerous people and institutions. We first thank Julia Obertreis, Kirsten Bönker, and Sven Grampp for organizing the December 2013 conference "European Television beyond the Iron Curtain" in Erlangen, Germany, where we met and decided to launch this project together. Sabina Mihelj and Anikô Imre were in Erlangen too, and they have provided invaluable help and encouragement ever since. Generous, multiyear funding from the Foundation for Baltic and East European Studies and the Riksbankens Jubileumsfond supported our research. Our own institutions, including the University of Wisconsin-Milwaukee (UWM)'s Department of History and Center for Twenty-First-Century Studies and Södertörn University's Department of Media and Communications Studies, provided crucial support, time, and opportunities to share our research at all the right moments. Colleagues at each of our institutions generously welcomed and included our international coauthor when we each spent a semester at the other's institution in 2016 and 2017, respectively. At Södertörn University, we especially thank Stina Bengtsson, Staffan Ericson, Johan Fornäs, Ekaterina Kalinina, Anne Kaun, Julia Velkova, and Patrik Åker. At UWM, we especially thank Tom Haigh, Jennifer Jordan, Elana Levine, Mike Newman, Rick Popp, and Jocelyn Szczepaniak-Gillece. Ferenc Hammer, Faith Hillis, Dana Mustata, Alexis Peri, Irena Reifova, Erik Scott, Victoria Smolkin, Susanne Wengle, and Espen Ytreberg provided important feedback and moral support.

In this book, we describe how producers of satellite broadcasts struggle to negotiate the liveness of broadcasting across many time zones. At times, we have found ourselves taking advantage of the fact that we live seven time zones apart, coordinating and synchronizing our workdays, since often when one of us was heading home for rest and family time, the other was just waking up. Still, writing across two continents has often been a challenge, and we warmly thank the organizers of the many conference panels, edited volumes, and special journal issues, whose interest in this project and thoughtful feedback allowed us to be in greater synch. We would like to extend our special thanks to the following colleagues for their support and feedback throughout the project: Andrew Jenks, Alice Lovejoy, Mari Pajala, Lisa Parks, James Schwoch, Asif Siddiqi, and the participants in "Networked Histories," the preconference that we organized as part of the International Communications Association's 2018 annual convention in Prague. We would also like to thank the scholars far and wide who have challenged our ideas, pointed us in the right direction, and had a profound influence on the final shape of this book. We thank Bohdan Shumylovych, for generously sharing his own research into Lviv's Earth station and providing crucial references, and Martin Collins in Washington, D.C., František Šebek in Prague, and Viktor Veshchunov in Moscow for sharing their experiences and insights with us. We thank Valentina Andreevna Nesterova for sharing her memories of her late husband, the Ukrainian artist Vladimir Nesterov, even as her country was under attack. We also are grateful to the many expert and generous archivists who made our work in numerous archives, presidential libraries, and other collections across two continents possible. We likewise thank the anonymous reviewers of this manuscript for their enormously helpful comments, and our editor at the MIT Press, Justin Kehoe, for his generosity and wisdom.

Finally, we thank our families, whose wonderful experiences getting to know each other and living abroad in Milwaukee and Stockholm while we worked out our ideas and traveled together to archives made this project meaningful personally as well as professionally.

INTRODUCTION

Shortly after Sputnik's launch in 1957, Hannah Arendt reported with alarm that "the immediate reaction, expressed on the spur of the moment, was relief about the first 'step toward escape from men's imprisonment to the earth.'"[1] Although, as Arendt wrote, "the man-made satellite was no moon or star, no heavenly body," Sputnik's ability to stay in the skies for even a short time seemed to suggest that this artificial creation of humanity had been "admitted tentatively" to the Moon's, planets', and stars' "sublime company."[2] The elliptical path of Sputnik meant that it continuously traveled across a new section of the Earth on each orbit, casting a global net, at least in theory. Amateurs and scientists across the world computed Sputnik's orbital track to be able to catch a glimpse of it or listen to the eerie beeps transmitted by Sputnik's radio transponder as the satellite passed by.[3] People on Earth tracked the presence, in the heavens, of an artificial object put there by other humans. Yet for Arendt and a generation of subsequent commentators, Sputnik initiated not chiefly the gaze up at space, but the gaze back down. She described with trepidation the possibility that humanity, now able to imagine itself truly capable of breaking Earth's bonds and regarding our planet from an external, Archimedean point, would become alienated from Earth, our only home.[4] For Arendt, our desire to imagine ourselves outside our planet was a disturbing shift—more significant, she argued, than the mastery of nuclear power, capable of destroying humanity entirely.

Arendt's response to Sputnik has served as a founding text for a richly theorized literature on the cultural and political impact of the view of Earth from space. Much has been written about the impact of two photographs in particular—"Earthrise," shot by the *Apollo* 8 crew on Christmas Eve of 1968, and the "Blue Marble" photograph taken by the *Apollo* 17 astronauts in December 1972.[5] Initial strong claims that these photographic views from space helped, for example, reenergize the environmental movement and peace activism during the Cold War have undergone subsequent examination and criticism.[6] Moving beyond the view of Earth alone, Lisa Messeri has argued that the launch of Sputnik and photographs of Earth from space contributed to "planetary imagination."[7] Satellite images, space exploration, and instruments such as the Hubble Telescope or Mars rovers, Messeri argues, allow us to think and conceptualize Earth and other planets in new ways, expressing a "desire to intimately know planets as worlds on which one can imagine being."[8]

One danger, however, of the forms of planetary imagination engendered by whole-Earth images from space is the elision of how partial the view from space really is, as well as how fragmented and unequally distributed space technology and infrastructure are on Earth. As Kelly Oliver has pointed out, "Blue Marble" and "Earthrise" produced the illusion of "whole earth," but they depicted only a fraction of the globe, reminding us that "the human perspective is always only partial."[9] Communications satellites offer an especially clear view of this inherent unevenness. They certainly operate in what Lisa Parks has called the vertical space, a field that "extends from beneath the earth's surface to the outer limits of orbit."[10] Yet, in contrast to the earthward gaze produced by "Blue Marble" and "Earthrise," the vertical mediation of communications satellites—which send and receive radio signals from one part of the Earth to another via satellites in space—do not pretend to produce a "whole earth," planetary imagination.[11] Satellite "footprints,"— the zone on the Earth's surface within which signals can theoretically be received—do not cover the whole Earth and are divided into sectors of stronger or weaker reception, defining zones of both access to and exclusion from audiovisual signal flows.[12] Falling within a satellite footprint, moreover, does not entail access to its signals. From the 1960s to the early 1980s, gaining access to satellite signals required the large antennae and signal-decoding technology of large, expensive, and sparsely scattered satellite Earth stations, as well as ground-based cable and radio infrastructure to redistribute the signal to the radio, telephone, and television. Communication satellite networks, in their early decades, thus most closely resembled other Earth-based communications and transportation networks, such as railroad, telegraph, and television systems, with their earthly terminals and excluded areas that undermined their promises of globality and instantaneity.[13]

Taking as its title Arendt's description of artificial Earth satellites as no heavenly bodies, this book explores the earthly history of the first two decades of satellite communications, tracing how satellite communications infrastructure was imagined, negotiated, and built across the Earth's surface, including across the Iron Curtain. The story of the US and European roles in the development and global spread of satellite communications has been well told, first by participants, insiders, and social scientists and more recently by historians of mass media and space technology.[14] We build on this work by incorporating the Soviet Union and other socialist countries into the story of how satellite communications technology was represented, built, and transformed into infrastructure in its first two decades. At the same time, we depart from the Cold War's binary, competitive framework that has, until recently, animated much of space historiography and telecommunications history. Instead, we focus on interaction, cooperation, and mutual influence across the Cold War divide.[15] By taking the expansion of satellite communications networks as a process of negotiation and interaction rather than a simple contest of technological and geopolitical prowess, we make visible the significant overlaps, shared imaginaries, points of contact and exchange, and negotiated settlements that determined the shape of the institutions and infrastructures for satellite communications in these formative decades.

THE EARTH IS LISTENING

Communications satellites' commercial function and entanglement with the history of television broadcasting have made them the banal other to the threatening world of secret defense satellites, or even to the environmental sensing satellites that have made the Earth programmable.[16] Precisely because of their quotidian status as media infrastructures, communications satellites represent the most earthly form of space technology, directing our attention toward a global history of space technology that belongs as much to the history of mass media as it does to the history of military technology, and that does not fit easily into a "Space Race" framework. While much of the study of "astroculture" has focused on rivalries and fantasies that take place off of

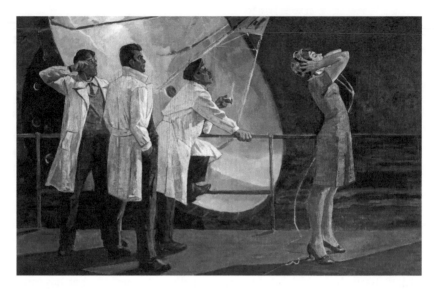

FIGURE 0.1
Vladimir Nesterov, *Zemlia slushaet* [The Earth is Listening], 1965. Reproduced with permission, Valentina Nesterova.

Earth's surface in space, communications satellites return us to the ground, emphasizing the importance of attending to the ways in which space activity reshapes the landscape and affects populations on Earth.[17] Finally, as a global medium from the beginning, communications satellites remind us of the fact that human space activity was transnational and global, regulated by and contested within international organizations like the United Nations (UN) and the International Telecommunication Union (ITU), long before the Apollo–Soyuz joint mission realized the goal of international cooperation in human space activity.

While communications satellites have arguably generated fewer utopian visions than human spaceflight, the early years of this new space-based medium did produce some striking efforts to visualize the promise of this new space technology. A 1965 painting, entitled *The Earth Is Listening* (figure 0.1), by the Soviet-Ukrainian painter Vladimir Nesterov, captures some of the main themes of these early satellite dreams and shows how very different the history of human space activity can look from the perspective of satellite communications. Here, a group of male scientists stands, in unusually chic and well-cut lab coats, in front of a large radio antenna (or "satellite dish"). They look up into the sky with a mix of wonder and authority, while before them stands a woman, dressed not in a lab coat but in a light, form-fitting dress and high-heeled pumps. She listens to some private, unknowable sounds via headphones, her eyes closed and her back arched in apparent ecstasy.

We could certainly interpret Nesterov's image as depicting a fantasy about the possibility of listening to space itself, like the radio amateurs tracking Sputnik in 1957. Or perhaps our listener is hearing some fantastical alien communiqué? The shadows on the satellite dish make it resemble the mottled surface of the Moon; the mounting hardware behind it likewise contains a series of small crescents. In both Soviet and American science fiction of the 1960s, satellite dishes frequently appear like giant ears, as part of space colonies and atop Moon rovers, as a representation of the expansion of human sensory power beyond Earth.

In fact, however, this is an image specifically about the pleasures and promise of communications satellites as infrastructures for earthly media. Nesterov's painting, based on study trips that he made to observatories in Kyiv and Crimea, celebrated the 1965 launch of Molniya-1, the Soviet satellite that marked the beginning of both satellite television transmissions within the Soviet Union and an extensive program of Franco-Soviet joint experimentation with satellite communications.[18] The sounds that are holding the listening woman rapt in Nesterov's painting are thus not from outer space. Rather, they are the sounds of other *people*, elsewhere on Earth itself, made audible by a proliferation of earthly infrastructures—antenna, headset, and cable snaking on the ground. The passing of the headset from the male scientist to the feminized media consumer, moreover, is a gendered account of the moment of infrastructural becoming, when new technology passes from the realm of scientific inquiry to that of everyday media consumption. Finally, Nesterov's painting suggests the powerful and sometimes unsettling intimacy of the world that satellite communications promised in its early decades. Rather than distant, alien landscapes to which humans dream of fleeing, Nesterov offers the embodied pleasures of listening (and of watching others listen) made possible by space technology that gazes back to Earth, rather than outward to the unknown. No heavenly bodies, communications satellites were launched and satellite ground stations built to wrap Earth in their infrastructural embrace.

SATELLITE TECHNOLOGY AS MEDIA INFRASTRUCTURE

This book tells the story of the global expansion of communications satellite technology as media infrastructures, which we define, following Lisa Parks and Nicole Starosielski, as "situated sociotechnical systems that are designed and configured to support the distribution of audiovisual signal traffic" and "material forms as well as discursive constructions."[19] Alongside the material and symbolic dimensions of satellite communications infrastructure, we emphasize the human institutions, networks, and interactions that shaped both the initial negotiation and construction and the ongoing maintenance of satellite communications infrastructure.[20] As we outline here, the chapters trace the initial fantasies and promises made about the potential of global satellite communications, follow the negotiation of a legal and political framework for regulating and governing global satellite communications, explore how satellite ground infrastructures were sold and built around the world, and demonstrate how satellite communications networks came to be increasingly integrated, as well as routinized and unexceptional, across Cold War political divides well before 1991.

The development of satellite communications may thus be understood as a process of becoming infrastructure—that is, of becoming embedded in other "structures, social arrangements and technologies," reaching beyond singular events or sites, and eventually fading into the background only to become "visible upon breakdown."[21] The claim that infrastructures are visible only when they fail is widespread and, as Brian Larkin notes, only partly true, since infrastructures can be anything "from unseen to grand spectacles and everything in between."[22] Satellite communications infrastructures, as we show in the following chapters, offer an especially striking example of this oscillation between poles, as they were at once hidden and remote and widely promoted, highly visible, and even spectacular.[23]

The development of satellite communications also offers a new perspective on the shared immanental nature of both infrastructure and media. The variable visibility of infrastructure points to the broader question, "When is an infrastructure?" posed by Susan Leigh Star and Karen Ruhleder in their classic essay "Steps toward an Ecology of Infrastructure." For Star and Ruhleder, the answer to this question is based on the way that infrastructures bridge local and global relationships; they contend that "an infrastructure occurs when the tension between local and global is resolved."[24] Tensions between

the local and the global, as well as the desire to resolve them, were especially pronounced in the case of satellite communications, since these were
planetary infrastructures that operated, from the beginning, across Cold War
political divides. While the full resolution of this tension is presumably never
fully achievable in practice, the aspiration toward its resolution reminds us
of a broader body of literature about the promises that infrastructures make,
including promising access to a global technological modernity.[25]

Yet we also suggest that media share a double status as both infrastructure
and event in themselves. As Joseph Vogl has argued, events are being communicated through media, while at the same time, the very act of communicating is an event in itself. As a result, Vogl suggests, media make things
visible, while in the process also displaying "a tendency to erase themselves,"
echoing Bowker and Star's account of infrastructures tending to "fade into the
woodwork."[26] The task of media studies, according to Vogl, is to acknowledge
the event character of media and how the process of becoming media is the
"coming together of heterogeneous elements—apparatuses, codes, symbolic
systems, forms of knowledge, specific practices, and aesthetic experiences."[27]
Drawing on Vogl's approach, *No Heavenly Bodies* offers an account of how
satellite communications technology became global media infrastructure
via the gradual integration and configuration of a similarly diverse array of
technologies, broadcast practices, institutions, symbolic representations, and
terrestrial networks.

The wide swath of "heterogenous elements" that came together to form
contemporary global satellite communications infrastructures have tended
to fall into the gaps between scholarly literatures on space technology and
media infrastructure. As James T. Andrews and Asif A. Siddiqi have argued,
the topic of space exploration "has generally attracted techno buffs or
political historians."[28] This may also be said of satellite communications,
where questions of ownership, control, and political power have dominated
research, especially during the early years of satellite development.[29] Institutional histories accompanied the 1964 creation of the US Communications
Satellite Corporation (COMSAT) and the International Telecommunications
Satellite Organization (Intelsat), the US-dominated, quasi-multilateral international satellite communications organization led by COMSAT as a majority shareholder. Intelsat's Soviet counterpart and rival, Intersputnik, also an
international organization modeled directly on Intelsat, has received far less
scholarly attention, but the very minimal existing work is similarly focused

on Intersputnik's political and institutional history.[30] The stories of US and Western European political actors in this process—diplomats, White House communications officials across several administrations, European state tele-communications officials, and public broadcasters—have been relatively well documented. Likewise, much of the early literature focused on legal prob-lems posed by the regulation of orbital space within existing frameworks of international law, outlining the positions of various states during such events as UN Educational, Scientific and Cultural Organization (UNESCO) confer-ences and the negotiations leading up to the 1967 Outer Space Treaty.[31]

More recently, scholars of space, communications, and US foreign policy have begun to challenge and deepen the analysis of the initial generation of institutional histories of Intelsat in particular. Historians have turned their attention to a broader set of US domestic and global actors and begun to document the frequently imperialist beliefs and strategies that underlay much of early satellite communications policymaking in the US, Europe, and the United Nations. David Whalen and Hugh R. Slotten have pointed out the ways in which the John F. Kennedy administration sought to strengthen US government control over satellite communications, displacing telecommu-nications corporations like AT&T that had invested heavily in satellite tech-nology.[32] Sarah Nelson has traced the role played by both US officials and UNESCO in excluding governments from the Global South from having a meaningful influence over the ways that satellite networks were built and operated, despite the latter's active efforts to improve the structural inequali-ties in media flows and infrastructures.[33] As Nelson shows, US officials moved to preempt more genuinely multilateral alternative models for global satellite communications governance. When research on the distribution of television content and the imbalance in communication flows began to be produced (most famously in a 1974 UNESCO report by Kaarle Nordenstreng and Tapio Varis), it chiefly served to document the very inequalities that officials from the Global South had protested more than a decade earlier.[34] By the time that research on global television flows intensified in the 1990s, however, satellites were no longer seen as a potential solution to the problem of unequal global media flows, and thus they were seldom discussed in their own right.[35]

Reconsidering the global expansion of communications satellites as media and communications infrastructures offers a new view of this technology and the actors and forces that shaped it. Considered alongside the nearly simul-taneous expansion of undersea fiberoptic cables, satellite communications networks appear both as part of the broader expansion of global media networks in this period and as an important outlier. Where cables were likewise invisible far under the sea, their urban-based infrastructure was built on the foundations of older media networks. By contrast, the Earth-based infrastructure of communications satellites had to be remote to minimize radio signal interference. While cable terminals, like the internet data centers that followed them, were often purposefully concealed by those who built and controlled them, satellite Earth stations were widely promoted, reflecting their status as key examples of Space Age architecture of the 1960s and 1970s that offered the countries who built them a symbolic connection to the Space Age's ultramodern technology.

At the same time, the rhetoric that accompanied the arrival and expansion of satellite communications infrastructures reminds us that they were often understood not through the lens of human activity in space—the vertical axis—but rather as a successor to horizontal media and transportation networks.[36] Like aspiring railroad and electricity monopolists before them, US officials proposed a "single global system"—that is, Intelsat—that could girdle the entire planet with American technology. As one early scholarly account of Intelsat's formation put it, Intelsat was "built by ambitious and pluralistic people who looked to the future. From their pioneering drive to the Western frontier and their building of the transcontinental railroad, to the determination to harness the atom and send a man to the moon, Americans have consistently taken up challenges with enthusiasm and all feasible resources."[37]

This enthusiasm for US leadership in the construction of global satellite communications infrastructure as the closure of the frontier found subtler guises as well. As Diana Lemberg has argued, US efforts to expand Intelsat globally were framed in terms of the promotion of the "free flow of information," an objective that served to justify and conceal the global expansion of US power.[38] At the same time, US officials, as well as television executives who sought to enhance the prestige of their own medium by linking to space, drew on longstanding utopian rhetorics of communication that promised to foster universal friendship and understanding through media circulation.[39] The development of satellite communications infrastructure was thus accompanied by both promises of national progress and modernity through Space Age technology and the erasure of politically problematic global interconnections and reinvigorated imperial relationships.[40]

As Lemberg, Parks, and others have noted, infrastructure's promises of technological modernization and global unification remained promises only, and indeed they served to conceal renewed imperialism and exclusionary, hierarchical political visions. Excellent new research in the history of US, Soviet, and European space activity has begun to focus on this relationship between space and globalization during what Alexander C. T. Geppert has called "the post-Apollo age": the ambivalent era of transition, retrenchment, and disillusionment with investment in space technology after 1969.[41] As Geppert has pointed out, communications satellites and their infrastructure offer the most obvious and concrete example of the entangled, causal relationship between human space activity and globalization.[42] Paul Edwards likewise identifies satellite systems, and specifically Intelsat, as a starting point for what he calls "infrastructural globalism."[43]

Exploring the relationship between space and globalization has led to the expansion of scholarly work documenting the resistance by, contributions of, and impact of space activity on people in the Global South, where much of Earth's essential launch and tracking equipment has historically been located. This work takes seriously the roles of officials, scientists, workers, and community members in what Asif Siddiqi has called "departure gates," the places on Earth from which humans access space, such as the European Space Research Organisation (ESRO) launch site in French Guiana and satellite tracking bases in Madagascar, Kenya, and elsewhere.[44] Centering the construction of space infrastructures on Earth in space history not only brings a wide range of new people, places, and views of the objectives of space activity into the picture, but it also suggests the impossibility of separating the history of human space activity into geopolitics versus dreams, or "cosmopolitics" versus "utopianism," as Andy Jenks describes this division.[45] Instead, as Pedro Ignacio Alonso has argued, "it is precisely the tension between peoples, sites and cosmic space . . . [that] compels us to stop considering elements in isolation."[46]

Despite the capacity of photography from outer space to produce and communicate the idea of Earth as a whole object, the scientific use of satellites and other sensing technologies has always produced knowledge about different Earth and near-Earth systems—the oceans, the Arctic, and the Ionosphere, to name a few.[47] Starting in the late 1950s, the Soviet Union constructed a network of satellite tracking stations stretching across a number of African countries just north of the equator to collect data about the world, not least through the practices of geodesy, the science of measuring the shape of the Earth, its gravity, and its orientation in space. As demonstrated by Siddiqi, these tracking stations "provided the knowledge required for Soviet scientists to propose a new model of the shape of the Earth, one that was an alternative to the conventionally agreed on US model," and thus quite literally reshaped the understanding of Earth.[48] While the following chapters may not explore the scientific production of multiple Earths, we find that by attending to the interaction and integration across the Iron Curtain, we can address how communications satellites allowed a reshaping of planetary conceptualizations.[49]

COLD WAR INTERACTION AND INTEGRATION ACROSS THE IRON CURTAIN

Although the focus on the post–Space Race era and the connections between space technology and globalization has expanded the geography of space research to include the Global South, the same process has not extended to the Soviet Union and the socialist world more generally. While Soviet space activity naturally occupies a central place in the historiography of human spaceflight and competition in space during the Cold War, with a few recent exceptions, the same interest has not extended to the Soviet role in the development of applications satellites.[50] Another major goal of this book is thus to make visible the significant role played by the Soviet Union in shaping contemporary communications satellite institutions and networks. We find that the Soviet role in the development of space communications infrastructure looks very different from our existing accounts of the Soviet space program as autarkic, secretive, and driven by a binary, competitive, Space Race framework. Instead, we demonstrate that satellite communications infrastructures that were not only entwined across borders but mutually constituted and shaped by extensive mutual interaction, mimicry, and shared understandings of how satellite communications should ideally be institutionalized.

Most broadly, international cooperation in communications satellites, with its commercial as well as scientific and political motives, was part of the process of economic globalization that accelerated during the years of détente. Since the commercial payoff was often somewhat distant or required significant investment before profits would be realized, communications satellite network infrastructure was built somewhat more slowly and haltingly than, for example, gas pipelines in Cold War Europe. Nonetheless, we argue, the institutionalization of international satellite communications networks and the construction of communications satellite infrastructure followed roughly the same pattern that European historians of technology have described as "hidden integration," in which technical networks such as gas pipelines connected European countries in ways that often directly undermined prevailing Cold War geopolitical logic.[51]

No Heavenly Bodies, therefore, locates the history of satellite communications within recent work on the neglected history of Eastern European socialist countries' extensive commercial and technical relationships with the postcolonial world.[52] As James Mark, Artemy M. Kalinovsky, and Steffi Marung point out, "the idea of Western capitalism as the only engine of globalization [has] bequeathed a distorted view of socialist and postcolonial states as inward looking, isolated, and cut off from global trends." These accounts, Mark et al. argue, "ignore not only the agency of so-called peripheries in the creation of global interconnection but also the possibility that interconnection between peripheries might be considered a form of globalization similar to the intensification of interaction between 'the West and the rest.'"[53]

Within this broader context of the socialist world's active role in processes of globalization and exchange with the developing world, satellite communications infrastructures in the 1960s–1980s, with their very high costs and monopoly on launch capacity, present a somewhat unique case. Excellent recent work on infrastructural integration and scientific exchange in Cold War Europe have tended to argue that infrastructures and scientific exchanges continued to expand across political borders largely without regard to the ups and downs of Cold War high politics.[54] Here, by contrast, we show how actions at the pinnacle of Cold War diplomacy in the early 1970s—the Nixon–Brezhnev summit of 1972—led, largely unintentionally, to the gradual but ultimately substantial material and institutional integration of the Intersputnik and Intelsat networks. Thus, while it is possible to see the mere existence and gradual expansion of Intersputnik as a Soviet-led satellite communications networks as a defeat for US policymakers, who sought a single network, we suggest that the competitive framework of triumph/ failure itself may be unsuitable for histories of global technopolitical systems. We propose instead to focus on unintended consequences, efforts to assuage ideological anxiety, and other dynamics that produced network integration, the fragmentation of Intelsat's initial monopoly, and new, planetary ways of thinking and acting.

We thus depart from the comparative approach to US and Soviet technical infrastructure taken by Benjamin Peters in his history of US and Soviet efforts to build what became the internet.[55] Rather than presenting Intersputnik's satellites chiefly as technically inferior to and less successful in global markets than Intelsat's during the first decades of satellite communications, we point out that they were simply different. Although early Soviet communications satellites indeed had lower capacity than Intelsat satellites, in terms of the number of telephone and television channels that they could offer, they also cost less. By taking a transnational approach that emphasizes interaction and mutual influence, moreover, we show how even asymmetrical, marginalized actors can reshape global networks and produce new planetary conceptualizations.

At the same time, we find that Soviet influence on the institutionalization of satellite communications and the construction of global ground infrastructures was predicated on the fact that Soviet officials' goals in these negotiations were not always significantly different from those of their Western European counterparts. Soviet space program officials wanted to benefit from the commercial opportunities created by this new technology. In the pursuit of a global satellite network, efforts to commercialize space were central, not marginal. Certainly, Soviet officials were happy to employ the rhetoric developed by nonaligned and postcolonial countries as they competed with the US for developing international clients. And they definitely sought to undercut Intelsat's prices and offer a more equitable revenue-sharing arrangement within Intersputnik. Nonetheless, Soviet officials' main objective was to recruit paying clients who would allow the Soviet Union to expand and profit from its own global satellite network.

SOURCES

Gabrielle Hecht has observed that the decision to center high-cost, state-controlled technologies like satellites risks limiting the frame of analysis to superpowers and other wealthy countries, "not because these systems don't extend elsewhere, but because the richness of metropolitan archives, the fascination with hegemonies, and the seduction of revealing the hidden politics

lurking in large systems all make it seem as though the most important stories remain grounded in the superpowers and in Europe."[56] This book does
not entirely escape this problem, not least because our research was cut short
by the COVID-19 pandemic and Russia's second invasion of Ukraine in February 2022. There is great need for further studies that focus on the experience of non-Western countries that participated in satellite communications
as users and Earth station owners. Nonetheless, we feel lucky to have been
able to conduct extensive, multiarchival research in Moscow, including the
archives of the Soviet Ministry of Communications, the Russian Academy of
Sciences (which holds the records of Interkosmos, the organization that coordinated Soviet international cooperative efforts in space), the Soviet Communist Party's Central Committee, the Intersputnik organization, and Soviet
Central Television.

While control over space-related decision-making was highly centralized
in the Soviet Union, and thus most archival holdings were in Moscow, places
far from Moscow were central to satellite infrastructure, institutions, and culture.[57] Vladimir Nesterov's *The Earth Is Listening,* for example, was painted by
an artist who was born, educated, and lived in Soviet Ukraine, and who based
his paintings of Soviet space infrastructure on sites that he visited within
Ukraine's borders. Like the chic scientists in his painting, Nesterov himself
wore a beret in the 1960s and for the rest of his life. His work thus reminds us
that Ukraine, along with Georgia, Kazakhstan, and other ostensibly peripheral Soviet territories, were in fact central to the construction of Soviet space
communications infrastructure and the production of new planetary ways
of thinking. We reached out to archivists and colleagues in Lviv and Kyiv
and visited the National Archive of Contemporary History of the Republic of Georgia in Tbilisi. We also noted moments where the remote places
where Earth stations were built asserted their presence in distant Moscow
archives. For example, the volatile, shifting permafrost soils in Siberia, prone
to destroying freshly poured Earth station foundations, offered their own
forms of resistance to both local engineers and central bureaucrats.

Since *No Heavenly Bodies* traces transnational influence and interaction,
however, our research could not be limited to post-Soviet archives. Moreover, the actions and influence of Soviet and socialist-world participants *do*
appear in US and European archival sources, both as participants and col-
laborators and as objects of American and European diplomats' and officials'

speculation, anxiety, and scorn. In largely passing over these moments when socialist world actors become visible in Western archives, historians and media scholars have tended to replicate the dismissive attitudes of US officials toward the socialist world displayed in these sources.[58] *No Heavenly Bodies* thus also incorporates archival sources from the US State Department and from several presidential library archives, reading them against the grain to identify moments where Soviet and Eastern European socialist countries' actions, both real and anticipated, shaped US and European decision-making or offered revealing glimpses of Western state actors' attitude toward their socialist-world counterparts. Finally, we incorporate archival sources from international broadcast organizations as well as state and corporate broadcasters, including the European Broadcasting Union (EBU), BBC, NBC, and Soviet Central Television, who were at the forefront of negotiating and airing international "satellite spectaculars" that crossed Cold War borders to promote the potential of this new technology.[59] Taken together, these multiperspectival sources offer us a view of the interactions of scientists, politicians, broadcasters, corporations, and satellite Earth station operators around the world in constituting new forms of transnational media governance and planetary thinking that emerged in the 1970s.[60]

OUTLINE OF THE BOOK

No Heavenly Bodies follow a roughly chronological structure, beginning with early efforts to imagine and promote satellite communications technology and tracing how satellite communications institutions and infrastructures went from being imagined and experimental to established and quotidian. Chapter 1, entitled "'Towers in the Sky': Satellites and Emerging Global Media Infrastructures," situates satellite broadcasting in a wider historical context, starting in the presatellite era, by outlining some of the early visions of global communication, from nineteenth-century fantasies to mid-twentieth-century broadcast technology experiments that aimed to cross the Atlantic. As the satellite era begins, the chapter describes the hopes and expectations for satellite communications, documenting the ways in which satellite broadcasting, even at the height of the Cold War, was always conceptualized as a trans–Iron Curtain endeavor. The very first Telstar experiments were planned to include not only the US and its Western allies, but also the Soviet Union.

Likewise, the Soviet Molniya network was to a large extent a national project, but it also had a transnational dimension through exchanges and technical cooperation between France and the Soviet Union beginning in 1965.

Chapter 2, "Promising Liveness: Contested Geography and Temporality in Live Satellite Broadcasting Events," traces how planned Franco–Soviet broadcast exchanges were overshadowed by a 1967 transnational effort to showcase the power and infrastructural promise of communications satellites—the BBC-led "Our World" broadcast in June 1967 and Soviet Central Television's "One Hour in the Life of the Motherland" in November of the same year. The chapter traces the evolution of plans for "Our World" over two years and demonstrates how BBC organizers' efforts to use the broadcast to present London as the center of a new kind imperial network in the age of decolonization did not go uncontested. Soviet, Polish, and Czechoslovak television officials involved in the planning process rejected the BBC's ideas about how to represent the spatial and temporal qualities of this new global network, ultimately reshaping the broadcast even after the socialist bloc's last-minute withdrawal. We then trace how the Soviet side's different understanding of the nature and purpose of live global presence was then realized in a second live satellite broadcast, "One Hour in the Life of the Motherland," that aired on Soviet Central Television just months after "Our World" and was directly based on the Soviet plans for the BBC broadcast. By comparing and contrasting these two satellite spectaculars, we make visible the contested spatial and temporal visions for what satellite broadcasting should mean and how competing ideas of what Lisa Parks has called "global presence" were at play in "Our World" and "One Hour in the Life of the Motherland." The chapter concludes by exploring the afterlife and significance of this largely forgotten socialist alternative account of what satellite-mediated "global presence" could or should accomplish.

Chapter 3, "Fragmented from the Beginning: The Entangled Origins of Intelsat and Intersputnik," moves away from public representations and goes behind the curtain of Cold War negotiations over the human and institutional infrastructures of satellite broadcasting. The chapter centers around the formation of the two main organizations in satellite communications during the Cold War, the US-led Intelsat in 1964, and the socialist and Soviet-led Intersputnik in 1968. Juxtaposing internal conversations and conflicts on both sides of the Cold War divide, we trace how the Soviet announcement, in 1968, that it would form its own satellite communications network,

Intersputnik, intersected with Western European partners' discontent with the US dominance of Intelsat to reshape the latter's permanent institutional structure from 1971 onward. We find that the mere existence of a Soviet alternative to Intelsat led the US to make greater concessions to both European and developing world members, creating an Intelsat structure that was much less able to prevent the formation of rival regional networks. Throughout these negotiations, both sides clearly understood that Intersputnik would be closely integrated with Intelsat, even if the Soviet Union did not officially become an Intelsat member until the late 1980s.

By the mid-1970s, the integration and interaction of Intelsat and Intersputnik on an organizational level were paralleled by a comparable process taking place in the realm of satellite ground infrastructure. In chapter 4, "'Space Begins on Earth': Selling, Building, and Representing Satellite Earth Stations," we recount how Intelsat and Intersputnik built a global network of satellite Earth stations, the large terrestrial antennae and control centers required to send and receive signals from space in the era before direct satellite broadcasting. As this process unfolded, both the US government and Western corporations engaged in building Intelsat Earth stations sought to link Earth station construction firmly to exclusive membership in one Cold War network. Focusing on the efforts to sell and build satellite communications ground structure around the world, as well as the technical features of early satellite Earth stations themselves, we demonstrate how these earthly networks threatened Cold War boundaries and complicated US efforts to maintain Intelsat's primacy. The chapter also explores how the new satellite Earth stations were represented in images designed to sell them to national telecommunications agencies around the world and celebrate their construction as evidence of their new operators' connections to modern space technology. At the same time, we show how images of satellite Earth stations also contributed to concealing the unsettling transnational, and trans–Iron Curtain, connections these new infrastructural buildings enabled.

Chapter 5, "Hotlines, Handshakes, and Satellite Earth Stations: Infrastructural Globalization and Cold War High Politics," traces how and why Intelsat's and Intersputnik's initially separate networks became integrated and layered over time, increasingly meeting Bowker and Star's definition of infrastructure as invisible except upon breakdown. We describe how the negotiation of a redundant and unused satellite backup to the cable hotline between Moscow and Washington as part of the Nixon-Brezhnev summit of

1972 led to the construction of Intelsat Earth stations in the Soviet Union and Intersputnik stations in the US and the gradual integration of the Soviet Union and several other socialist states into Intelsat via the processes and institutions for maintaining and updating Earth stations, as well as via the desire for global satellite-enabled broadcasting for media events on both sides of the Iron Curtain.

The book concludes with an epilogue that connects the history of satellite communications infrastructures to the current, privatized space economy and to contemporary debates and fears about the expansion of global communication infrastructure. Tracing the evolution of Intelsat and Intersputnik through the privatization of the 1980s and 1990s, we show how the socialist world's consistent pursuit of profit in space technology helped advance space communications privatization. Finally, we revisit current debates about space communications infrastructure, locating them within a longer trajectory of beliefs, hopes, and fears about global media networks.

1 "TOWERS IN THE SKY": SATELLITES AND EMERGING GLOBAL MEDIA INFRASTRUCTURES

On April 14, 1961, a motorcade carrying the Soviet cosmonaut Yuri Gagarin was heading toward Moscow and Red Square. He was seated in the first limousine together with his wife and Nikita Khrushchev, with live television images covering their journey and displaying the crowds gathered along the road, cheering and waving flags as they passed. To television viewers in Sweden, and other Western European countries, it was difficult to see what was really happening in Moscow. The quality of the picture was poor, and every now and then it was lost and replaced by a gray, flickering screen. During one of these interruptions, Sven Wahlström, a Swedish commentator and expert on the Soviet Union, was asked whether the doves seen earlier in the broadcast were released on this festive occasion. Wahlström explained that they were a permanent feature of the Moscow cityscape, and continued with the voice of the confident expert: "Doves actually have a protected status in Russia. You are fined, if I remember correctly, fifteen Rubles if you hit a dove with your car and kill it. They are fully protected." The story about doves in Moscow is a typical feature of the live broadcast. While the agreement between Soviet Central Television and national broadcasters in Western Europe was that they were obliged to follow the Russian commentary verbatim, the constant interruptions demanded some improvisation to present viewers with something more than just a silent, gray screen.[1] As the motorcade sped on, the commentary noted, "Well, maybe now, we may have the picture back. Something went wrong before [the signal reached] Helsinki." The rest of the broadcast went smoothly.[1]

Gagarin's return to Moscow was the first live broadcast from the Soviet Union to reach television viewers across Western Europe. No satellites were involved in the relay, which instead used the two terrestrial television networks in Europe, Intervision and Eurovision, linked. It was little more than a year before the first satellite broadcast across the Atlantic using the American satellite Telstar, but this brief episode in broadcast history illustrates some important features of transnational broadcasting, features that would also define satellite broadcasting as it was introduced over the following decade.

The launch of Sputnik and Yuri Gagarin's flight, of course, were important landmarks of the Space Race and widely considered to be significant victories for the Soviet Union during the Cold War.[2] However, the act of broadcasting Gagarin's return to Moscow to audiences in Western Europe was not an example of unidirectional propaganda, but rather a cooperative effort by broadcasters in both the Soviet Union and Europe.[3] While this particular broadcast was something of a last-minute solution and a surprise, it was also just the latest example of the longstanding ambition to turn television into a global medium.[4] And while the remark about the status of doves in Moscow could be read as empty chitchat, it also reflected a recurring rhetoric that presented global television as being able to bring people together and promote understanding between citizens around the world. For television to serve as a tool for global understanding and friendship, however, all parties had to agree to the terms of a transnational broadcast, ranging from negotiations over programs to be produced and shared, to, as in the earlier example, whether local commentary could deviate from the script. Finally, shared visions and ambitions and successful negotiations and cooperation also needed to be realized by technology carrying the television signal over vast distances. Infrastructures had to be built and maintained, and the links had to be reliable to avoid situations wherein viewers had to be informed that "something went wrong before . . . Helsinki."

This chapter traces the history of global broadcasting just before and after the arrival of satellite broadcasting technology in light of the questions raised here. We begin by outlining the dreams and ambitions of global communication before the advent of satellite communications, setting the stage for the early satellite spectaculars that will be the focus of chapter 2. Next, we turn to the broadcast organizations and the negotiations that provided the framework for transnational program exchanges in the years just before the advent of satellite broadcasting. The final part of the chapter narrates the early years of communications satellites, examining early experimental broadcasting networks using American Telstar and Soviet Molniya satellites that would later develop into the large-scale satellite networks Intersputnik and Intelsat and mapping early infrastructural developments by looking at satellite launches and ground station construction.

VISIONS OF GLOBAL COMMUNICATION

There is a famous cartoon published in *Punch* magazine in December 1878, showing a girl, Beatrice, in Ceylon speaking to her parents in London using "Edison's telephonoscope," a device capable of transmitting "light as well as sound."[5] The image shows Beatrice in a tropical setting and her parents watching her on a large screen in their cozy London home, with Beatrice updating them on newcomers to the colony. The cartoon is sometimes referred to as an example of how the invention of the telephone soon fostered fantasies of television, seeing at a distance.[6] A decade later, in 1892, *Punch* published a science fiction–like satire with a story about future interplanetary communication, "Reading the Stars a la Mode."[7] The story was based on an extract from a notebook of "the Secretary of the Earth and Mars Intercommunication Company," starting with an entry in a future 1899 noting that volcanoes and oceans have been identified on Mars, and that "Marsians [sic] were trying to speak to us. They seemed to be making signals." After making contact and establishing communication, humans are abruptly brushed aside in 1927, with the Martians replying to a request for communication, "Don't bother; can't attend to you just now. We are talking with the planet Jupiter." The fictional notebook ends with an entry from 1934, noticing that the "London, Jupiter, Venus, Mars, and North Saturn Aerial Railway Company" was now in operation. Joshua Nall has pointed out that the *Punch* story about interplanetary communication should be seen in light of the establishment of a so-called New Astronomy and the development of more powerful space observatories, telegraph networks, and immediate news distribution.[8] Together, these developments served as a catalyst for ideas about the annihilation of time and space that were prevalent at the time.

New communication technologies sometimes provide an opportunity to reassess distance. The items published in *Punch* in the late nineteenth century illustrate two fantasies of communication across vast distances. Beatrice's call to her parents in London foreshadowed a future in which global communication would be about distance, connectivity, and empire, spurring discussions about the Earth as a global village. The vision of communication with Mars pushes the limits of communication even further, expanding into outer space. These two works create a link between related discourses, global communications networks on Earth, and the broader idea of space communication and exploration.

These imaginations were global and planetary, establishing a vision of communication as either encircling the globe or reaching well beyond it. Yet when we turn to the implementation of new broadcasting systems, it soon becomes evident that they were far from global in reach. Prior to the introduction of communications satellites, media audiences were at best transnational rather than genuinely global. In this sense, with regard to media before the satellite age, the term "global" refers to an idea, a technological imaginary, or a fantasy of modernization.[9] Leaving the science fictions of the late nineteenth century behind and turning to mid-twentieth-century experiments and prototypes for global communication, we instead encounter transnational infrastructures, communications networks that reach across national borders but are limited to regional, or at best transatlantic communication. In short, whereas "global communication" must always refer to technological *imaginaries*, "transnational communication infrastructures" refers to the actual practices of broadcasting, which were never truly global.

The *Punch* cartoon has often been read as the ideation of television, of communicating the idea and formulating a hypothetical solution to audiovisual communication across vast distances.[10] But it also reveals another important driving force behind the development of satellite communications—telephony. Since the first undersea transatlantic telegraph cable in 1866, there had been continuous efforts to increase the transmission quality and the capacity of these cables to be able to carry not only telegraph messages but also voice conversations that demanded a greater bandwidth and signal strength. In 1956, just one year ahead of Sputnik, the first telephone cable across the Atlantic, TAT-1, was inaugurated, with a capacity of fifty-two telephone channels.[11] While broadcasting remained the most spectacular and eye-catching use of communications satellites, it was the prospect of expanding the market for long-distance telephony, together with data transfer, facsimile, and other technologies, that made the idea of a single global communication system commercially viable.

TRANSNATIONAL EXPERIMENTS

Already during the first postwar decades, a wide array of ideas and technologies were introduced in pursuit of the global transmission of television broadcasts. Some of these technologies explicitly aimed, albeit unsuccessfully, to achieve what James Schwoch has called the "holy grail of the transatlantic crossing."[12] In 1945, for example, Westinghouse started to experiment with an airborne broadcasting system called Stratovision, which would place television transmitters in airplanes to cover fourteen large city areas in the US.[13] The basic idea was to elevate the transmitter to bypass the problem of line-of-sight communication, which limited broadcast towers on Earth. By using airplanes, Stratovision aimed to cover areas up to eighteen times larger than the existing broadcast towers. And while it was envisioned as a national system, it soon inspired a similar international system intended to carry television broadcasts, as well as facsimile, business information, and other data. According to RCA chair David Sarnoff, its rival system, Ultrafax, would outperform Stratovision, stressing not only television capabilities but also its strategic importance and military uses since its reach was intended to be worldwide. Rather than airplanes circulating metropolitan areas nationally, Ultrafax would include "an airborne radio relay system (that) could serve as a constant watchman to intercept guided missiles that might be traveling in our direction."[14] Introduced to President Harry S. Truman, and the US Congress in 1948, it has been described as a "wireless internet of sorts for mid-century—[that] combined radio relays with high speed film processing such that it could send not only documents and sound but also film to remote locations."[15] The most expansive and elaborate idea for global broadcasting was a system called UNITEL, proposed in the early 1950s. One of its main components was the North Atlantic Relay Communication (NARCOM) system, which, using a network of microwave relay towers to bridge the Atlantic Ocean, "strategically placed on mountaintops, islands, and rimming the oceans of the world."[16]

NARCOM aimed to connect the US and Canada with Europe, ultimately linking North American broadcasters to the European Broadcasting Union (EBU) and its Eurovision network. NARCOM would employ already-existing coaxial cable routes on the North American side, from New York City, over Quebec and onward to St. Lawrence, and from that point using ultrahigh frequency (UHF) relay towers to carry the signal to southern Greenland and on to

Iceland. The transmission path between Iceland and the Faeroe Islands is the longest jump of the entire route, where the signal has to travel approximately 290 miles, a far longer distance than could be managed using line-of-sight communication. Beyond-horizon communication, therefore, was needed.[17]

While never installed, NARCOM highlighted one key problem in transatlantic broadcasting, the need for extremely tall broadcast towers to overcome vast distances. None of these visions, Stratovision, Ultrafax, or UNITEL, was ever fully realized, but they all suggest how seriously plans for global television networks were pursued—plans that would later be realized by the introduction of communications satellites, which finally created "towers" high enough to enable transatlantic broadcasting. At the same time, these failed precedents demonstrated that communications satellites alone were insufficient without terrestrial communications networks.

THE BEGINNINGS OF TRANSNATIONAL BROADCASTING

Radio waves defy national borders. Already early in the history of radio, this prompted broadcasters, and particularly the BBC, to embrace what Simon J. Potter called "wireless internationalism," expressed by, for instance, the BBC's motto "Nation Shall Speak Peace unto Nation."[18] But the fact of inherently transborder radio broadcasting also led to calls for an international regime of radio broadcast regulation already from the very first years of radio broadcasting. This was particularly pressing in Europe, where national broadcasters often shared the same broadcasting space and frequency allocation was necessary to avoid interference.

The case of television was somewhat different, however. The reach of television signals was more limited, even though there were areas of cross-border broadcasting, particularly in densely populated regions in Europe.[19] The incentive for international cooperation in television instead focused on establishing technical standards and legal frameworks, as well as on enabling television program exchanges between national broadcasters.[20]

The first examples of cross-border television broadcasting were bilateral, beginning with the so-called Calais experiments during the summer of 1950. The Franco-British exchanges continued to develop, and in July 1952, eighteen programs were successfully broadcast from Paris to London during the so-called Paris week. A year later, in May and June 1953, fifteen programs were broadcast from London to Paris. This series of broadcasts culminated with the coronation of Queen Elizabeth II on June 2, often mentioned as a milestone in television history.[21] The coronation was broadcast live in Great Britain and France, but it also reached television viewers in Germany, Netherlands, and Denmark. The network of coaxial cables that enabled the broadcast to be seen simultaneously in five countries may be regarded as the foundation of the television networks that later would be established on the continent: Eurovision and Intervision.

The European Broadcasting Union (EBU) and the International Organization for Radio and Television (OIRT) were children of the Cold War, born in the aftermath of World War II when their predecessor, the International Broadcasting Union (IBU), was split in two. During the interwar period, national broadcasters across Europe gathered under the same organizational umbrella of the IBU.[22] At the time, radio broadcasting was not part of the agenda of the International Telecommunication Union (ITU), which then was mainly focused on telegraphy.[23] As radio broadcasting proliferated across Europe, the need for international regulation became obvious. European broadcasters suffered severely from interference caused by radio waves on shared frequencies crossing national borders, prompting ten national broadcasters in Europe to establish the IBU in 1925.[24]

Recognizing the problem of interference, the IBU immediately set out to address the use of wavelengths and frequency allocation. The distribution of wavelengths between twenty-nine European countries was decided by the so-called Geneva plan, which was accepted in 1926.[25] The work on frequency allocation and monitoring the radio spectrum was thus a key field of operation for IBU. From the beginning, the IBU had a dual mission. On the one hand, it decided on frequency allocation to protect the airwaves of national member organizations; on the other hand, it was instrumental in organizing program exchange and other activities that would enable a more transnational radio landscape in Europe.[26]

Immediately after World War II, the Soviet Union proposed a new international broadcast organization, the International Broadcasting Organization (OIR), arguing that the IBU was too closely affiliated with the Nazi regime in Germany during the war. The negotiations that followed were an indication of Cold War tensions to come. The French advocated the need to transform IBU into a truly international organization, modeling it upon existing organizations such as the ITU and the United Nations (UN). The British, instead, wanted the IBU to remain the same, and they were very concerned that allowing the Soviet-allied Eastern European countries to join the IBU under a UN-like voting scheme would damage the balance of power within the IBU, as had already happened, from the British perspective, in the United Nations. Other Western European members soon joined the British in their critique of changes to the voting rights. While the IBU survived for a few more years, several member-countries resigned from the IBU, leaving it seriously weakened.[27] By 1950, the IBU was dissolved after a conference in Torquay in February, where the EBU was established by an initiative of the BBC. As a consequence, broadcasters in Western and Eastern Europe had two separate and distinct broadcast organizations beginning in 1950: the EBU and OIR.[28]

Despite this bifurcation into separate organizations, the EBU and OIR shared a similar structure. Both had a legal committee, program committee, and technical committee, and both operated in similar ways. Most important, they both engaged in program exchanges within their networks, establishing live transmission networks, Intervision (OIRT) and Eurovision (EBU), in 1960 and 1954, respectively.[29] When Intervision was announced in February 1960, the network was limited to Czechoslovakia, Poland, Hungary, and the German Democratic Republic (GDR), but the press release indicated that the network would expand in the near future to include the Soviet Union, Bulgaria, and Romania.[30]

Intervision was announced just days ahead of a joint meeting between the EBU and OIRT in early February 1960, with program exchange between the two divided networks, in radio as well as television broadcasting, as an important point on the agenda. The meeting outlined three main types of exchange: television recordings, news exchange, and direct exchanges (i.e., live transnational transmissions). While direct exchanges were the least common of the three, the live transmission network would later be crucial to the planning of future live broadcasts by satellite. In a separate meeting of the respective technical committees, EBU and OIRT officials agreed that the organizations should exchange maps of current and future nodes in the networks, and the OIRT even proposed five junction points where the two networks could be linked. This suggestion was a bit premature to the EBU, which immediately referred it back to national authorities to decide.[31] Still, in early 1960, there were ongoing negotiations and a mutual understanding between the EBU and OIRT regarding program exchanges between national broadcasters in Western and Eastern Europe. In addition, the networks carrying the television signals were operational and covered large parts of Europe. All that remained was to link Eurovision and Intervision into a continent-wide network.

The Gagarin broadcast in 1961 represented the first successful attempt to link Eurovision and Intervision. But the television images from Moscow had to follow a somewhat unexpected route on their way to television viewers in both Eastern and Western Europe. Since the Soviet Union was still not linked to Intervision, there was no clear path to relay the broadcast from the Soviet Union to the other members of the OIRT, such as the GDR or Czechoslovakia. Instead, a temporary link was established between Tallinn in Soviet Estonia and Helsinki in Finland by shifting the direction of a transmitter in Tallinn and mounting an ordinary rooftop antenna on top of a water tower in Helsinki.[32] Using the already-existing cable linking Helsinki and Stockholm, the signal was fed further into the Eurovision network, and from there to the Intervision network of the OIRT countries.

At the time, in spring 1961, the Soviet Union was not part of the Intervision network. As a consequence of the temporal link between Tallinn and Helsinki, not only did Soviet Central Television manage to organize a live broadcast to Western Europe, but a link to national broadcasters in Eastern Europe was established as well. The temporary link was established by cooperation between Soviet Central Television, Yleisradio in Finland, and the BBC. While the broadcast of Gagarin's return to Moscow was an ad hoc solution, the plans of a live broadcast from Moscow to London had been in the making for several years. At the time the world learned about Gagarin's spaceflight, a delegation from the BBC was already in Moscow preparing for a live transnational broadcast from the upcoming May Day parade a couple of weeks later.

While the Gagarin broadcast was a spectacular example of a media event, the history of transnational broadcasting in the early 1960s was often somewhat more quotidian and focused on meeting broadcasters' practical needs.[33] Already prior to the Gagarin broadcast, there had been program exchanges using the Eurovision and Intervision networks, such as during the Rome Olympics in the summer of 1960.[34] While there were a number of converging incentives for program exchanges—cultural, political, and technical—the overriding reason for program exchanges, also between OIRT and EBU, was financial.[35] National broadcasters across Europe were all struggling to fill their schedules, a problem particularly urgent for smaller broadcasters with scarce resources for producing their own programming. One solution to this was to engage in bilateral or multilateral program exchanges, in which Intervision and Eurovision played a crucial role. Notably, the flow of television programs enabled by the two networks was rather one-sided, with a large influx of programing from Eurovision to Intervision.[36]

Yet even this rather unbalanced flow of programming contributed to the construction of relationships across national and ideological borders. Alongside multilateral exchanges enabled by the networks, there was bilateral cooperation between broadcasters, sometimes using Intervision and Eurovision as a relay. These broadcasts, which included the exchange of recorded material in addition to live transmissions, contributed to the creation of bilateral relationships that could foster further plans. Between April 1961 and July 1963, the BBC received about a dozen programs relayed via Eurovision, including media events, such as celebrations of the cosmonauts Gagarin and Gherman Titov, political events like the May Day Parade and a speech by Khrushchev, as well as live sporting events. In the same period, the BBC also recorded six programs in the Soviet Union, starting with a football match between Aston Villa and Dynamo Moscow, a performance of *Romeo and Juliet* from the Bolshoi Theater, two performances by the Moscow State Circus, and *World Zoos: Moscow*, presented by David Attenborough.[37]

The Attenborough *Zoo* program demonstrated especially clearly how broadcast cooperation could lead to the development of professional relationships over time. In February 1958, Attenborough presented the idea of traveling to Soviet Central Asia in search of animals such as the Siberian tiger, snow leopard, ibex, and Marco Polo's sheep, "a very impressive horned sheep," for an episode of *Zoo Quest*, his first major nature documentary series. Attenborough ended his program pitch by saying, "We should hope to bring back some of the smaller animals we captured for presentation to the London Zoo."[38] In a March 1958 letter to the Soviet State Committee for Television and Radio Broadcasting (Gosteleradio), BBC officials assured their Soviet counterparts that none of the strictly protected animals would be captured, but they hoped to bring some of the "smaller, less rare, but nonetheless interesting and unusual creatures" to London.[39] The proposal was met with little enthusiasm. Vasilii Evgeniev, the head of international relations for Gosteleradio, replied with a short letter noting that such an expedition could not be organized, and as for the animals, he suggested approaching the Soviet Export Organization.[40]

Over the coming years, the BBC repeated its proposal several times, without luck. However, despite the apparently fruitlessness of this correspondence, several of the BBC personnel involved in the *Zoo Quest* negotiations were later involved in other, successful recorded as well as live program exchanges with the Soviet Union.[41] The episode of the rejected *Zoo Quest* proposal in 1958, moreover, was by no means the first contacts between the BBC and Soviet Central Television. Three years earlier, a Soviet delegation had visited the UK, a visit that was reciprocated in 1956 when a delegation from the BBC spent twelve days in late April and early May in Moscow, Kyiv, and Leningrad. The visits aimed not only to explore possible program exchanges but also to facilitate the sharing of expertise and technical knowhow.[42] Visits such as these, like meetings of broadcast professionals within the framework of the EBU and OIRT, forged personal relations and provided the groundwork for more extensive cooperation in live satellite broadcasts a decade later.[43]

These personal connections between broadcasting personnel were accompanied by efforts at the OIRT and EBU to build the administrative structures needed to make program exchanges and cooperation more routine. The technical committees of the OIRT and EBU met for the first time in Helsinki in 1957, and again three years later in Geneva, to prepare for transnational exchanges via such actions as putting protocols in place for the exchange of technical information. The main point on the agenda for the 1960 meeting was to discuss "Operational questions concerning the mutual exchange of television programmes," including possible junctions where the two networks could be linked. The technical committees also exchanged codes of practices and coordination of pre-transmission test schedules to synchronize their procedures. The 1960 meeting also followed up on some of the agreements from three years earlier, such as the exchange of information regarding broadcasting stations and coordination of measurements of ionospheric propagation carried out by the two organizations.[44]

These efforts to facilitate exchanges using the terrestrial networks later proved crucial to the use of communications satellites for broadcasting. Live satellite broadcasts, beginning with the Telstar experiments addressed next, were not the product of a brand-new infrastructure. Of course, the satellites themselves and the space-to-Earth link created by ground stations were remarkable newcomers to the world of global communication. But to relay the signal to national broadcasters and television viewers, broadcasters needed to connect the new space infrastructures to existing ground infrastructures for television, such as Intervision and Eurovision in Europe and the radio relay and cable networks of the US. In short, the spectacular satellite broadcasts of the 1960s relied on already-established networks of live broadcasting.

At the time of the introduction of communications satellites in the early 1960s, a number of factors paving the way were already in place. The vision of transatlantic exchange and coproduction between the US and the UK were firmly established, while perhaps not always recognized in broadcast histories, as Michele Hilmes has argued.[45] In Europe, the obstacle to overcome was not the great distances and oceans, but the political and cultural differences among national broadcasters. The joint efforts by European broadcasters to bring the EBU and OIRT closer created an organizational and institutional framework for transnational broadcasting, even if the final decisions regarding cooperation and exchange remained with national broadcasters. These broadcasters in turn, as with the BBC and Soviet Central Television, added another layer of personal relationships among television professionals, who often shared ambitions and ideas regarding transnational program exchanges. Earlier bilateral and EBU-OIRT collaborative efforts, from the late 1950s onward, led to the construction and regulation of networks for live broadcast exchanges, allowing the transnational distribution of broadcasts. In the summer of 1962, on the eve of satellite broadcasting, the ground and institutional infrastructures for transnational broadcasting were already well developed.

"TOWERS IN THE SKY": COMMUNICATION SATELLITES, ORBITS, AND FOOTPRINTS

What remained to be solved was the problem of building a radio tower tall enough to facilitate transatlantic broadcasting—the task at which NARCOM, UNITEL, and other early transnational broadcast projects had failed. For communications officials and firms, the arrival of communications satellites was the long-sought answer to this problem. An advertising brochure by Bell Telephone Systems, promoting the Telstar project in the early 1960s, explained how satellite technology would at last solve these problems, conquering the obstacle that the curvature of the Earth posed to the vision of a global communications network. As the Bell leaflet explained, microwaves travel in a straight line and therefore cannot follow the curvature of the Earth

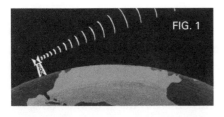

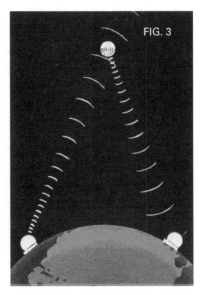

FIGURE 1.1
Towers in the Sky, Bell Telephone System, "Project Telstar," n.d. Reproduced with permission, Nokia Corporation and AT&T Archives.

when communicating over vast distances. Consequently, argues the leaflet, a successful global communications network by means of micro relay, crossing the Atlantic Ocean, would demand a tower mid-ocean, 475 miles high. The solution to this problem, Bell's writers explained, is "a special kind of 'tower' . . . —a tower in the sky—a satellite."[46] The promotional brochure depicted satellite broadcasting as easy to understand, a natural development and extension of overland radio broadcasting that relied upon "a few simple ideas and many complicated details."[47] Among these details were the mobility of the tower of the sky, the orbit of the satellite, and the footprint produced by its coverage (figure 1.1).

Artificial satellites can orbit the Earth via low-, medium-, or high-Earth orbits. The three orbits are suitable for different kinds of satellites and uses, with Earth-observing satellites using low-Earth orbits and weather-monitoring satellites using high-Earth orbits. Since the speed of the satellite is determined by the gravity of the Earth, satellites in near-Earth orbits have faster revolutions around the Earth, whereas satellites in high-Earth orbits have slower revolutions. In addition to the distance to Earth, the satellite's inclination and the eccentricity of the orbit are important. The inclination is the angle of the orbit in relation to the equator, whereas the eccentricity is a measure of the shape of the orbit. A low eccentricity describes a circular orbit, while high eccentricity describes a highly elliptical orbit. The high-Earth orbit is located 42,164 kilometers above the center of the Earth (roughly 36,000 kilometers above the surface), and a satellite in this orbit has the same rotation speed as the Earth and thus remains in the same position above the surface of the Earth, while some drift may occur. Therefore, if a satellite is placed in the high-Earth orbit, and also directly above the equator, it will remain in what is called a "geostationary orbit"—that is, it will remain relatively stationary relative to positions on Earth, allowing satellite antennae on Earth to point at the satellite continuously, with only minor adjustments needed.

Satellites in a medium-Earth orbit may either be placed in a semi-geosynchronous orbit, each orbit lasting twelve hours and with a very low eccentricity (close to zero). The other option is to place the satellite in a highly elliptical orbit in which the Earth is close to one edge. As a consequence of the shifting distance to Earth, the velocity of the satellite varies as it travels its orbit: when close to the Earth, the satellites move quickly, only to slow significantly as the distance to Earth grows due to the elliptical shape of the orbit. These basic characteristics of satellite orbits had to be taken into consideration in the design of satellite communications systems. Determining the characteristics of potentially useful orbits was, of course, a central issue in space research.

The first experimental communications satellites used elliptical orbits, which meant that the satellite was usable for broadcast transmission only during the period of time when the satellite's footprint passed over the ground stations of a given system. As discussed further next, the American Telstar satellite used only four dedicated ground stations to communicate with Earth, and the Soviet Molniya satellites used existing Soviet ground stations that were chiefly used for space science rather than communications. Both the US and Soviet experiments, however, were limited and struggled to adapt existing technologies to the needs of communications satellites. This limitation was further accentuated by the fact that each side had only one communications satellite in operation at any one time, meaning that switching between two or more satellites to allow more continuous broadcasts was impossible. Whether they were using Telstar or Molniya as a vehicle for relaying television programming, broadcasters faced a similar problem, in that the satellites allowed only a brief window of transmission as the satellites traveled across the area where the ground stations were located. The use of highly elliptical orbital paths thus meant that the footprint of the satellite moved across the surface of the Earth and the link between satellite and Earth was operational for only a limited time.

There were two means of addressing this problem. By placing the satellite in a geosynchronous or geostationary orbit, its footprint would remain over the same terrestrial area, allowing continuous use of the satellite.[48] In theory, then, three geosynchronous satellites would provide almost global coverage.[49] Alternatively, placing multiple satellites in elliptical orbits and dedicating ground stations to track their movements and continually picking up the signals as they passed across the sky would allow continuous broadcasts even while the footprint moves across the Earth. This use of elliptical orbits and tracking stations on Earth had been previously employed in space research, such as when mapping the ionosphere.[50]

The US and its Western allies settled for a geosynchronous orbit using satellites in high-Earth orbit. The footprint of these satellites would cover a large portion of the world, and just as important, would allow continuous broadcasting twenty-four hours a day. With the Molniya system, on the other hand, the Soviet Union chose the other path—multiple satellites in highly elliptical orbit—which better suited their specific geographical location and needs.[51] The geosynchronous orbit with the satellite placed directly above the equator was unsuitable for the Soviet Union for several reasons. First, it was only with great difficulty that the Soviet Union even could place a satellite in orbit above the equator due to the nonequatorial location of its launch sites. To succeed, the Soviet space program had to settle for a satellite of a mere 100 kilograms, a payload far too small to provide the capacity of the communications satellite that was needed.[52] The geostationary orbit above the equator also had the disadvantage, from a Soviet point of view, of not reaching the far-north Arctic regions.

Based on the location of the launching site, the Soviet engineers decided that the satellite would be placed in an orbit with a 65-degree inclination and high eccentricity. The highly elliptical orbit would allow the satellite to have an apogee (i.e., the farthest distance from Earth) of about 40,000 kilometers above the Northern Hemisphere, while only 400 kilometers (perigee) when passing the Southern Hemisphere. The elliptical orbit meant that the satellite would travel at its highest speed over the Southern Hemisphere, while spending a lot more time when covering the Northern Hemisphere. As a result, the twelve-hour revolution around the Earth allowed between eight and nine hours of continuous broadcast between Moscow and Vladivostok. The choice of this orbit thus solved the problem of not only how to launch the satellite into the proper orbit from Soviet territory, but also how to secure coverage for the geographically vast country. Finally, the orbit allowed the launch of a satellite with a payload up to 1,600 kilograms, far more than what would have been possible if trying to reach a geostationary orbit above the equator, regardless of launch site location.[53]

The various orbital paths used by Soviet Molniya satellites and what would later be the Intelsat system, however, did not pose an especially great obstacle to linking the two systems for transnational broadcasting. Instead, as will be explored in later chapters, the two systems used different radio frequencies for communicating between the satellite and the ground stations, requiring costly adjustments at, for example, Goonhilly Downs in the UK, to be able to receive broadcast signals from Molniya satellites.[54] Even this obstacle, however, was far from insurmountable.

TELSTAR AND EARLY TRANSNATIONAL SATELLITE BROADCASTS IN EUROPE

The story of Telstar, the first dedicated communications satellite, launched from Cape Canaveral in Florida on July 10, 1962, and operational for about six months—its functional life cut short by radiation from nuclear weapons testing—is well known.[55] As James Schwoch notes, "the roots of Telstar lie in long-distance and intercontinental telephony," with American Telephone and Telegraph (AT&T) seeking, for commercial purposes, to increase the voice circuit capacity and quality across the Atlantic.[56] AT&T had developed and manufactured Telstar by an agreement with the National Aeronautics and Space Administration (NASA), but the US government was concerned early on that AT&T would create a global monopoly similar to the situation in US telecommunications. As a result, the John F. Kennedy administration passed the 1962 Communications Satellite Act, which established a new company, the Communications Satellite Corporation (COMSAT), with the purpose of setting up a single global satellite communications system.[57] AT&T was the largest shareholder in COMSAT, however, and remained influential in the development of Intelsat, as discussed in chapter 3. Similarly, while television captured the audience's attention, telephony was a key source for revenue for Intelsat in the years ahead, as voice circuit capacity increased continuously.[58]

The logistics of the first transatlantic broadcasts, which Telstar made possible, were quite challenging and fragile. Each of Telstar's elliptical orbits around the Earth was (and still is, albeit as space junk rather than an operating satellite) about 158 minutes, but due to its specific alignment, Telstar's footprint only covered the Earth stations in Andover, Maine (US), Goonhilly Downs (UK), and Pleumeur-Bodou (France) on certain orbits, and even then for only a brief time (figure 1.2).[59]

62043

FIGURE 1.2
Ground stations during Telstar experiments, Bell Telephone System, "Project Telstar," n.d. Reproduced with permission, Nokia Corporation and AT&T Archives.

During Telstar's sixth orbit, testing of the satellite's communication capacity began. These tests included relay of data, facsimile, and news copy, and a phone conversation between AT&T's Fred Kappel and Vice President Lyndon B. Johnson. Kappel introduced himself and announced that he was calling from the Earth station in Andover, Maine, and then the conversation continued:

KAPPEL: The call is being relayed through our Telstar satellite as I am sure you know. How do you hear me?

JOHNSON: You're coming through nicely, Mr. Kappel.

KAPPEL: Well, that's wonderful.[60]

The conversation was later covered by television and relayed by Telstar on its seventh orbit, and it was soon confirmed that the televised images relayed by Telstar had been received in France at the Earth station in Pleumeur-Bodou. Even the experimental phase of satellite communications was designed not merely for testing, but also for broadcast and promotion, showcasing the power of this new technology.[61]

At the same time, from Telstar's earliest, experimental orbits, competition over access and control of international broadcasts was already apparent. Due to miscalibrated equipment, the reception at Goonhilly Downs was weak and the British could only briefly discern flickering Telstar images. When the tests continued on the next day, both the French and the British relayed television programming back to the US using Telstar. Whereas the material coming out of Goonhilly Downs was primarily technical, such as test patterns, the French seized the opportunity to promote French culture and entertainment. Rather than test patterns, the French relayed across the Atlantic a recording of Yves Montand singing "La Chansonette" after being introduced by the minister of postal services and telecommunications in France, Jacques Marette. In response, the British and other EBU members accused the French of exploiting the experimental broadcasts to promote their own national interests.[62] A few days after this supposed "coup," the president of the EBU, Olof Rydbeck, reassured EBU members that the dispute between Pleumeur-Bodou and Goonhilly Downs was a misunderstanding, and all parties should look forward to the television exchanges planned to take place on July 23, 1962.[63] The dispute over these first experimental broadcasts, however, suggested that international tensions over access, representation, and control over satellite broadcasts, including tensions within Cold War alliance blocs, arose simultaneously with communications satellites themselves.

Once the Telstar broadcasts moved beyond their initial experimental phase, US and European broadcasters sought to engage more regional and national broadcasters as participants. The programming during what was called the "America to Europe" and "Europe to America" broadcasts on July 23, 1962, included contributions from across the American and European continents.[64] Both broadcasts emphasized the vast distances covered by the broadcast, featuring images from the US-Canadian and US-Mexican borders, and ranging from San Francisco to New York. The European broadcast was introduced by Richard Dimbleby of the BBC, noting that all images would be live from a total of fifty-four cameras across Europe. Here as well, contributions spanned the continent, from Gällivare in northern Sweden to large cities in continental Europe: Vienna, Rome, Belgrade, Paris, London, and others. Throughout the broadcast, the local presenters continuously emphasized the vast distances covered and the large audiences reached.[65] Nonetheless, in these America-to-Europe and Europe-to-America broadcasts, the only Eastern European participant was Jugosłowiańskie Radio i Telewizja (JRT) of Yugoslavia, a full member of the EBU since its founding.

However, in the months leading up to the broadcast, there had in fact been a number of overtures to the Soviet Union to take part. In early June, Aubrey Singer, assistant head of outside broadcasts at the BBC and executive producer of EBU Satellite Programme, went to Moscow to meet with Konstantin Kuzakov, vice chairman of the USSR State Committee for Broadcasting, to discuss Soviet participation in the Europe-to-America broadcast.[66] Writing from his hotel upon leaving Moscow on June 3, 1962, Singer sent Kuzakov a letter outlining a mutual understanding that Soviet Central Television was prepared to take full part in the exchange and to transmit on its full network both the European and American contributions. The agreement also said that no statements by heads of state were to be broadcast if they were deemed by Soviet Central Television to be of a political nature. Finally, the letter noted that if the Soviet Union decided not to include the American broadcast, the EBU would not "be able to offer its program to Soviet Television or accept any Soviet contributions."[67] Despite these clear disagreements over the inclusion of political leaders in the broadcast, the proposal for Soviet participation led to the development of a detailed script, opening with a greeting in front of the Kremlin, followed by live broadcasts from St. Basil's Cathedral, from inside the Metro station, and finally from a football game inside the Lenin stadium (today Luzhniki Stadium).[68] These images, however, never reached the audience across Europe and the US.

The clause about not including any heads of state in the broadcast reflected Soviet anxiety about allowing the US president to speak directly to Soviet viewers. US officials, by contrast, were eager to include a presidential greeting in the broadcast and unwilling to see that address removed from the broadcast by local broadcast organizations. While Singer met with Kuzakov in Moscow, Robert Mayer Evans of the US Information Agency wrote to Edward W. Ploman at the administrative office of the EBU to explain that the US position was that it would "seem natural for experiments on an American satellite to have a brief appearance by the American President," and suggesting that if any objections arose, Ploman should explain to Soviet representatives that "in Europe you have no control over the American program."[69]

A couple of weeks before the launch of Telstar, the Soviet participation in the broadcast became increasingly uncertain.[70] On June 27, 1962, Singer received a telegram from Kuzakov, confirming continued interest in the broadcast and noting that Soviet Central Television would send preliminary versions of a Soviet contribution.[71] However, the following day, a number of phone conversations and telegrams were exchanged, indicating that the cooperation was in jeopardy.[72] Less than a week later, the president of the EBU, Olof Rydbeck, sent a telegram to the chairman of the State Committee of Radio and Television, Mikhail Kharlamov, in which he put a definitive end to the plans of cooperation since he had not yet received a reply to a previous letter and noting that for technical reasons, they could no longer make changes.[73]

The events of July 1962 demonstrated that live television across the oceans was no longer a thing of the future, but viewers in Europe and the US experienced the beginnings of a global network of satellite television. The launch of the satellite was highly spectacular, of course, but just as important was the network of Earth stations that sent and received the signals to and from Earth. The four Earth stations carrying the Telstar broadcast emerged as the embryo of the future Intelsat network, which would carry the "Our World" broadcast five years later. While the cooperation with the Soviet Union failed, the efforts to include images from the Soviet Union in the "Europe-to-America" broadcast of 1962 prefigure plans to include the Soviet Union and other East European broadcasters in the "Our World" broadcast in 1967. As we argue in subsequent chapters, however, efforts to incorporate Soviet and Eastern European broadcasters into US- and European-led satellite broadcasts often had unintended consequences.

LAUNCHING MOLNIYA

Plans to include the Soviet Union in the July 1962 Telstar broadcasts focused on overland radio relay links, featuring an elaborate pathway that would route television signals from Moscow to Cologne via eleven nodes, including Leningrad, Warsaw, and Dresden.[74] A satellite link was not yet possible since the first successful Soviet communications satellite, Molniya 1, was not launched until three years later, on April 23, 1965.[75] The day after that launch, a satellite link was established between Moscow and Vladivostok, traversing the Soviet Union's vast territory. The first public broadcasts were conducted during the May holidays, initially on May 1 and then on May 9, the twentieth anniversary of the Soviet victory in the Great Patriotic War. While the first Telstar experiments to a large degree celebrated the technical feat of the broadcast itself, emphasizing the vast distances covered and the cultural diversity of the participating countries, the early Soviet broadcasts were employed to enhance the televised celebration of the main Soviet calendrical and revolutionary holidays; the initial May Day broadcasts were followed by satellite links connected to the revolutionary anniversary programming of November 6–8, 1965.[76]

Although the Molniya system was initially used as a national satellite broadcasting system—indeed, it was the first domestic satellite broadcasting system in the world—it too was transnational in its conception and use from the beginning. Just a few weeks after the November 1965 holiday broadcasts, a successful broadcast between the Soviet Union and France was carried out under a bilateral scientific agreement regarding experiments in color television.[77] The scope of this cooperation was initially limited to experimental transmissions because of the French commitment to the emerging Intelsat system. There were technical constraints on the cooperation as well; regular broadcasts and exchanges between the Soviet Union and France necessitated the installation of an additional antenna at Pleumeur-Bodou.[78] Nonetheless, experimental broadcast cooperation was renewed in 1966, when the French press reported on an agreement between France and the Soviet Union to "begin exchange by fall 1967 of 12 hours of color-television broadcasting per week between France and USSR via Soviet tele-communications satellite Molniya."[79] US diplomats noted that the Molniya system did not yet have the capacity to relay sound and images due to bandwidth problems, but they were now being addressed in cooperation with the French, and the new antennae being built at Pleumeur-Bodou would be used for the exchange. As international cooperation with the French unfolded, and as the Soviet Union participated in plans for what became "Our World," a circumglobal satellite broadcast planned for June 1967, Soviet telecommunications officials oversaw the construction of an extensive dedicated ground station network to support domestic and international satellite broadcasting, and especially satellite signal distribution across the Soviet Union's more remote regions, which were not and could not easily be connected to Moscow by radio relay networks.

Parallel to the experimental use of Molniya satellites for television broadcasts, plans were made to construct a dedicated system of communications satellite ground stations, relaying television broadcasts and telephone communications. It has been noted that in 1960, over 80 percent of international telephone traffic originated or terminated in the US, illustrating the importance and dominance of US actors such as AT&T to the development of communications satellites as a means to increase the quality and capacity of voice circuits across the Atlantic.[80] In its early phases, Molniya was not incorporated into a system of international telephony, but the vast distances of the Soviet Union naturally added to the need of using communications satellites for telephony. The first Molniya satellite could be used for broadcasting one television channel, or alternatively forty to sixty channels of telegraph or telephone communications.[81] Just as with COMSAT and Intelsat, the Molniya system was continuously developed to be used for television and telephony, and the increased payload and use of the superhigh frequency band allowed Molniya 2 to simultaneously broadcast television and carry multichannel telephony.[82]

The network of ground stations, called Orbita, would cover a substantially larger portion of the country, with a planned network of twenty stations in operation by 1967.[83] The new Orbita ground network replaced an ad hoc Molniya ground system based on the existing Saturn ground network, which was used for a range of tracking, mission command, and scientific research applications.[84] Soviet communications satellite ground infrastructure was thus markedly domestic in its first two years. Yet domestic ground stations were equally available for global use, making possible Soviet live participation in a global satellite broadcast like "Our World," as discussed in chapter 2.

Representations in the Soviet press of the new Orbita ground station network depicted it as exclusively domestic, with all stations nestled, as in the Pravda graphics shown in figure 1.3, within crisply demarcated borders. However, plans for the international expansion of Soviet satellite communications networks were already underway even before the first Molniya satellite's launch in 1965. Nikolai Mesiatsev, the chair of Gosteleradio, described the origins of the international Soviet satellite communications network, Intersputnik, as the product of a conversation with one of the founders of the Soviet space program, the rocket scientist Sergei Korolev, before the Molniya system was up and running. Mesiatsev claims that he was first introduced to Korolev at a reception in the Kremlin by none other than Yuri Gagarin himself, whom Mesiatsev had gotten to know when they both were involved in the Soviet–Cuban friendship society. When Mesiatsev and Korolev met again, it was at a meeting of the Military Production Committee in the late fall of 1964, where Mesiatsev raised the specter of both the US formation of Intelsat, which would "encompass the whole world," and direct broadcast satellites, which would, he argued, "create a revolution in [people's] minds."[85]

Talking with Mesiatsev afterward, Korolev suggested that, in addition to working through UN organizations to regulate and limit direct satellite broadcasting when it became technically feasible, the Soviet Union should take its own steps by developing plans for its own global satellite broadcast network. Korolev, Mesiatsev recalled, sent him a note, sketched "so clearly" on four pages ripped from a "student's graph paper notebook" that it was easy to understand "without any special effort."[86] These pages laid out a system based on the launch of three, presumably geostationary, communications satellites in positions that would allow nearly global coverage. Mesiatsev's account of his exchange with Korolev reaffirmed the central premises of Korolev's cult in post–Soviet Russian space historiography, emphasizing Korolev's humility and generosity with his time, his ability to plan and foresee future technical possibilities far in advance (no Soviet geostationary satellites were launched until a decade later), and his unique ability to communicate complex scientific ideas with a clarity that made them comprehensible to nonspecialists.[87]

Карта размещения пунктов «Орбиты».

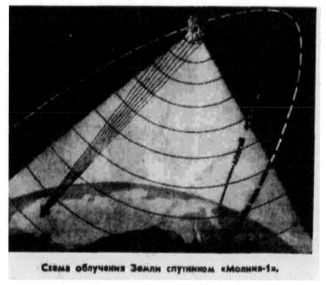

Схема облучения Земли спутником «Молния-1».

FIGURE 1.3
"Map of the Location of Orbita Stations" and "Diagram of Molnia-1's Coverage of the
Earth," *Pravda*, October 29, 1967, 3.

Yet what stands out here is that Soviet plans for their satellite communications network were conceived from the very start as global in scope, not chiefly because of Soviet ambitions to influence global audiences directly, but rather because they felt obliged to react to and counterbalance the power of the US-led network, Intelsat, newly created in 1964. In addition to the desire to "contain" a US-led telecommunications network, moreover, we might also explain Korolev's and Mesiatsev's planetary ambitions for Soviet communications satellites by pointing out that experiences and norms of transnational exchange and cooperation were well established among Cold War scientists and engineers on both sides of the Iron Curtain by 1964. Along with extensive Soviet engagement with countries of the Global South, these experiences encouraged and legitimized global and planetary thinking and the expectation that scientific advances and new technical infrastructures would be shared globally.

Mesiatsev's memories link the origin story of the Soviet international satellite communications network, Intersputnik, to the nationalist historiography of the Space Race, with its focus on individual "founding fathers" and emphasis on tit-for-tat rivalry between the US and the Soviet Union. Yet, much like the longer history of presatellite and early satellite transnational broadcast experiments that preceded the formation of either global satellite network, this particular arena for Space Age competition was highly multilateral, engaging European broadcasters and states on both sides of the Iron Curtain from the beginning and focused particularly on the construction of ground infrastructures, without which there could be no transnational or global broadcasting by satellite.

CONCLUSION

The early years of communications satellite technology thus remind us to attend to the wide range of human actors and places engaged in the expansion of satellite communications around the globe. Like the handful of experimental transnational broadcasts that took place just before and after satellite broadcast technology became available, early communications satellites themselves could not meet the definition of infrastructure proposed by Bowker and Star. Short-lived and unreliable, they were quite unable to "fade into the woodwork."[88] Unpredictable, unreliable, and without any backup redundancies in the system, their ability to support signal traffic as invisible, reliable "towers in the sky" was highly questionable.[89] The earthly networks that would support and further distribute satellite signals to terrestrial communications networks on the ground were equally patchy, incomplete, and unreliable. Yet just as remarkably consistent as breakdowns and failures were US, European, and Soviet broadcasters' consistent ambitions and efforts to use broadcast technologies to distribute television signals across Cold War borders. Those early visions played an essential role in both driving infrastructural construction and network integration and in defining and publicizing this new communications technology to global audiences from Europe to America. When the BBC and EBU announced plans for a spectacular first—a live satellite broadcast transversing the Northern Hemisphere and including the Soviet Union and its Eastern European allies—they were building, if not on an existing infrastructure, then on longstanding infrastructural visions and established human and institutional relationships.

2 PROMISING LIVENESS: CONTESTED GEOGRAPHY AND TEMPORALITY IN LIVE SATELLITE BROADCASTING EVENTS

For the BBC, European Broadcasting Union (EBU), and Soviet Central Television journalists and administrators, the goal of producing the first global satellite television broadcast sprang from their own long experience engaging with one another in various other transborder broadcasting efforts. These began, as we argue in chapter 1, long before television transmission by communications satellite was even a possibility. For producers at both the BBC and Soviet Central Television, moreover, employing space technology to continue these longstanding cooperative efforts offered a way to enhance their own prestige and institutional authority, connecting public service broadcasting institutions that were facing competition from either commercial competitors (in the case of the BBC) or more-established state media institutions like film studios (in the case of Soviet Central Television) to the prestige and utopian hopes associated with human space activity.[1]

The desire, among public service broadcasters, to create from whole cloth a satellite television event that would express and, momentarily, realize the promises of satellite communications technology for global television audiences, however, posed significant logistical, political, and creative demands. In this chapter, we trace the negotiations leading up to "Our World," a 1967 broadcast organized by the BBC in collaboration with public television services across the Northern Hemisphere, and analyze the *two* live satellite broadcasts that resulted from this heavily contested planning process. These included, of course, the "Our World" broadcast, which aired in June 1967 without the participation of the Soviet Union and its allies, who withdrew just days before the broadcast, and a little-known Soviet counterpart, entitled

"One Hour in the Life of the Motherland," which aired a few months later, in November 1967. As public service broadcasters worked to make visible one of the main promises of satellite communications infrastructure—the experience of global televisual liveness that Lisa Parks has called global presence—they had to navigate conflicting understandings, based on longer traditions of imperial and colonizing visual representations of space, of what global televisual liveness should mean to audiences.[2] The resulting broadcasts reflected both the underlying conflicts between Soviet and British understandings of how communications satellite infrastructure should reorganize global space, and a broader underlying similarity in planetary thinking and shared imperializing objectives. Moreover, both broadcasts' claims to historicity were undermined by the fact that satellite communications technology ultimately promised only to extend or accelerate existing media globalization processes rather than to break from Earth's surface or the past.

In the winter of 1967–1968, the famous, and famously self-regarding, Soviet television journalist Yuri Fokin looked back with pleasure at the Soviet Union's first live, transcontinental satellite broadcast, entitled "One Hour in the Life of the Motherland." The broadcast had crisscrossed the Soviet Union's eleven time zones with live satellite linkups from Tashkent, Leningrad, Tbilisi, Lviv, and many more Soviet television centers that were newly connected via the Soviet Union's Molniya communications satellite network, the first domestic satellite broadcasting system in the world. The broadcast celebrated the initiation of regular satellite broadcasting via Molniya and was timed to coincide with the weeks of festive broadcasting on Soviet Central Television that marked the fiftieth anniversary of the October Revolution. Satellite broadcasting was, the technical feat suggested, the latest step forward in the Soviet conquest of distance and progress toward global, revolutionary synchronization.[3] At the same time, the Soviet communications satellite system, as the product of enormous state investment, required explicit celebration to make it visible to and appreciable by audiences.

Fokin's reminiscences about the program focused not on evaluating the broadcast—his essay was accompanied by a variety of short reviews of segments of "One Hour in the Life of the Motherland" by other television critics. Instead, he described where the idea for the broadcast had come from. "It was the middle of August," Fokin's narrative began:

Students studying to be television commentators and correspondents were finishing their summer internships at Central Television. Vacation lay ahead and everyone's spirits were high . . . Eighteen young, energetic people who have recently started on the difficult path of television journalism—that's a lot of power . . . Various ideas and doubts passed through my head, and yet . . . Guys, what if we try to make a show, built around mini-reports? A show built around live segments from different ends of the country?

In just a few hours, Fokin claimed, "we had written a first draft of a scenario for the future broadcast called 'One hour in the life of the Motherland.'"[4] Fokin's account of the origins of "One Hour in the Life of the Motherland" presented it as the profoundly original brainchild of enthusiastic young Soviet journalists—a newly ascendant professional group that had pressed for and gained public prominence beginning during the years of rapid social and political change under Soviet premier Nikita Khrushchev.[5]

Fokin's origin story for "One Hour in the Life of the Motherland," however, was the purest fiction. In fact, "One Hour in the Life of the Motherland" emerged from a dense palimpsest of Soviet media precedents, from Stalin-era radio *pereklichki*, or "live link-ups," in which workers in factories across the Soviet Union reported on their success and challenged other far-flung collectives, to two high-profile Soviet journalistic projects, created in 1935 and 1961, that sought to document a single day in the life of the whole planet.[6] Soviet cultural production was in fact littered with earlier projects that aimed to both visualize and transcend the scale and diversity of Soviet territory, as well as that of the whole, potentially revolutionary planet.[7]

Yet the most immediate precursor for "One Hour in the Life of the Motherland" was not a previous Soviet documentary project, but the much better known satellite spectacular that had aired just months earlier, on June 25, 1967. That broadcast was the BBC-led "Our World," which was to include eighteen countries as well as regional broadcast organizations such as the EBU and the International Organization of Radio and Television (OIRT).[8] The Soviet Union had planned to participate in "Our World" until its withdrawal just days before the June 25 broadcast, chiefly in response to the outbreak of the Six-Day War between Israel and Arab states in early June 1967.[9] Fokin's narration of "One Hour in the Life of the Motherland" as a spontaneous new idea by young Soviet journalists thus required the complete renarration of the broadcast's complex origins, obscuring the long history of cooperation and negotiation with the West, as well as the longer

traditions of Western *and* Soviet globalizing and imperializing visual culture on which both "One Hour in the Life of the Motherland" and "Our
World" rested.

BBC producers were quick to denounce "One Hour in the Life of the Motherland" as a poor-quality, unacknowledged imitation of "Our World," as we
discuss next. But a similar process of forgetting and concealment characterized the broadcast of "Our World," in which the Soviet withdrawal was discussed briefly at the beginning and attributed to Soviet isolationism (rather
than to the Soviet government's antiwar position). Not evident in the final
broadcast were the years of joint organizational efforts between Soviet Central Television and the BBC, which exposed both conflicts in the organizers'
understandings of how global space and time should be interpreted and
represented for viewers and substantial shared beliefs about how satellite
communications technology presented an opportunity to assert a new imperial organization of space in the postcolonial world. Similarly invisible were
all the ways in which Soviet participation and pressure shaped the ultimate
broadcast of "Our World," despite the Soviet bloc's last-minute withdrawal.
In this chapter, we excavate the multiple contexts and entangled, trans–Iron
Curtain origins of the efforts of "Our World" and "One Hour in the Life of
the Motherland" to promote not only the potential of satellite communications technology to forge an international, if not truly global, television audience but also a mode of planetary thinking, emergent in this area, in which
shared human experiences of measuring and making meaning from the orbit
and rotation of the Earth in relation to the Sun were deployed in the service
of geopolitical and ideological objectives.

INFRASTRUCTURAL PROMISES: GLOBAL PRESENCE AND THE
GEOGRAPHIES OF LIVENESS

One of the most prominent discourses used to describe what satellite communications technology would mean for broadcasters and audiences was
that of televisual liveness. Western and Soviet theories of the mediated experience of live presence, what the Soviet avant garde filmmaker Dziga Vertov
and later Soviet television critics, following Vertov, deemed "the effect of
presence" [*effekt prisutstvie*], long predated the 1960s but were quickly seized
upon by broadcasters as especially relevant for television via communications satellite.[10] In the promotional material for "Our World," Aubrey Singer,

the program's chief editor, mixing claims about spectacularity drawn from circus promoters with the language of 1960s social and artistic undergrounds, described it as the "greatest show from Earth" and "a global happening by means of television."[11] Accentuating the program's specific temporality, inviting audiences to take part in an event unfolding live in front of their eyes, Singer's ambition echoed a well-established account of television as essentially live and immediate, which guided television production practices in this period.[12]

What made "Our World" unique as a live television event, however, was its (partially unfulfilled) global ambition. Unlike the extremely local spectacles of the circus show or happening, "Our World" would make the live televised image travel 200,000 kilometers, encircling the entire Northern Hemisphere. Lisa Parks has characterized "Our World" as promising viewers an experience of "global presence," which she describes as a Western fantasy predicated on a neocolonial vision of global geography and on the exclusion of countries of the Global South, which could not participate in or even watch "Our World," and were thus unable to experience the satellite-enabled global presence that the broadcast promised. As Parks suggests, the temporal and spatial claims that underlie the rhetoric of "liveness" are inseparable.[13]

An examination of the planning and production process of "Our World" and "One Hour in the Life of the Motherland" exposes at least three forms of spatial, temporal, and infrastructural inequality and rivalry implicated in the creation of a global live satellite broadcast, which we call "geographies of liveness."[14] First, despite the implicit claim of live productions that liveness is accessed and experienced equally across both time and space—in fact, even within the group of countries that contributed to and participated in the production of "Our World"—access to material infrastructures necessary for live broadcasts and their symbolic power was unequally distributed across spatial and geopolitical boundaries.[15] Second, these inequalities were subject to ongoing contestation as part of the broadcast production process. In the case of "Our World," these inequalities were *not* successfully effaced and made invisible by the producers in charge—instead, they were the subject of ongoing conflict and negotiation among the participating sides and were reflected in the broadcast's final form. At the same time, the material infrastructures created for the broadcast were largely invisible to the show's audiences and, indeed, sometimes secret.[16] As a result, the broadcast event could be promoted and celebrated in entirely contradictory ways in different

national contexts, as broadcasters and other participants presented conflicting accounts of their own places within the larger, global infrastructural networks that the broadcast created and celebrated. This surprising openness to
multiple claims and interpretations is a third feature of the geographies of
liveness underlying live, transnational broadcast events.

The process of planning and organizing "Our World" exposed all these
challenges because by necessity it engaged many participants—the powerful
US and Soviet space programs, which alone could launch communications
satellites, as well as many other countries whose national infrastructures of
cables and rebroadcast towers were essential to the program's success. Infrastructural negotiations quickly became questions of power and symbolic
representation that took spatial as well as temporal form. Who could claim
to be the center of the spectacular, modern, transnational infrastructure created to evoke the instantaneous quality of "Our World"? Who would be rel-
egated to its periphery, visually, within the broadcast, and, more materially,
in the actual infrastructure constructed for "Our World"? Pulling off this live,
instantaneous transcendence of space required the creation, over years, of an
elaborate set of plans, scripts, and technical networks—requiring, ultimately,
the creation of a command central that would take the lead in the negotiation of the broadcast's final form, represent itself as administering the broadcast during the show itself, and serve as the final authority in coordinating
technical decisions. Where would this controlling center of the live network
be located? Whose national infrastructures and personnel made the live temporality of "Our World" possible? These questions of center and periphery
lay at the heart of the broadcast's production of liveness. They were also, we
find, highly dependent on the actual construction of satellite infrastructure
in the participating countries, as well as the hotly contested symbolic representation of that infrastructure in the planning process.

COMPETING TEMPORALITIES AND GEOGRAPHIES IN THE PLANS FOR "OUR WORLD"

The planning of "Our World" started in 1965 with the ambition to "circum-
navigate the northern hemisphere" by midsummer's day (June 21) 1966. The
broadcast was conceived, from its earliest planning, as a trans–Iron Curtain
initiative, with Soviet participation central to the planning process. The earliest drafts of the show proposed live segments from various places in Europe,

the Soviet Union, Japan, and the US. In a letter dated December 9, 1965, Aubrey Singer, head of television outside broadcasts at the BBC and chief editor of "Our World," reached out to the Soviet foreign correspondent Henry Trofimenko, asking whether the Soviet Union would be interested in participating in such a broadcast.[17]

From the early stages of planning the broadcast's name, as well as the timing and network construction of the overall project, raised the possibility of conflict. Singer cycled through a number of possible titles for the broadcast. His initial provisional title for the broadcast was "The Longest Day— The Longest Way," linking the broadcast to the planetary temporality of the Northern Hemisphere's summer solstice, which he hoped to select as the date for the broadcast. "The Longest Way," in turn, suggested the effortful, even gratuitous transversal of planetary space via satellite. Before settling on "Our World," Singer proposed yet another temporally infused name: "Around the World in 80 Minutes," an obvious reference to the novel by Jules Verne. The temporal perspective communicated by these early program names, however, was not one of immediacy, liveness, and presence, but rather of calendar and clock time. Rather than the instant transmission of images, these proposed titles emphasized the longest day of the calendar (notably, in the Northern Hemisphere only—the Southern Hemisphere was entirely excluded) or the clock time of the program itself. In the end, however, the broadcast's organizers reverted to "Our World," a title stripped of temporal signifiers while still alluding to the interconnectedness of an imagined global village.[18]

The shift in title appears to be partly due to a conflict that quickly emerged between the BBC and Soviet Central Television over the program's proposed date. Soviet Central Television's leadership was quite critical of the chosen date since, from the Soviet perspective, June 21 was significant not chiefly as midsummer's day, but as the day before the anniversary of Adolf Hitler's invasion of the Soviet Union on June 22, 1941.[19] On March 7, 1966, following a meeting with Soviet Central Television executive Anatolii Bogomolov, Singer sent a note to him suggesting that perhaps the Soviet side could reconsider its objections to June 21 as a date for the transnational broadcast. He wrote that "after all, this program will be very much in the interests of peace, and on sad anniversaries of this sort surely constructive thinking would be welcome."[20] Of course, Singer's desire to retain June 21 as a date was driven by practical concerns: it was incredibly difficult to coordinate a program time and date with so many national services and international organizations. Yet the choice was also ideological. In 1966, the planned broadcast was organized around cosmic themes, and the BBC had selected June 21 precisely for its solar significance as the longest day of the year—a fact indifferent to violent and unequal human histories, which Singer sought to downplay in favor of a focus on planetary (or at least Northern Hemispheric) temporality and concerns. This episode exposed the difficulties of navigating both sides' different understandings of time and temporality. To the BBC, the cosmological, cyclical time of the solstice rendered the Soviet objections insignificant, while to Soviet Central Television, teleological, historical time—in which Soviet heroism and suffering in World War II was a crucial plot point—proved far more important. In this instance, the Soviet side eventually prevailed, and, following a postponement of the entire broadcast to the summer of 1967, the broadcast date was changed to June 25.[21]

The question of how precisely to visualize the planned broadcast's feat generated even more pointed conflicts between the broadcast's BBC organizers and Soviet bloc participants. Despite the regular recurrence of broadcaster and critical enthusiasm about the ostensible power of live broadcasting, liveness is not inherently interesting to viewers, nor is it even discernible without specific cues that distinguish a live broadcast from a prerecorded one. Live broadcasts thus must make their liveness visible, often by placing reporters outside so that weather, time of day, the clothing or appearance of passersby, and other details reinforce both the broadcast's liveness and, in the case of a multisite live broadcast, the transversal of distance. Early drafts of the broadcast referred to this effortful display explicitly, emphasizing the different times of day during which segments in different parts of the globe took place and thus stressing the importance of clock time to the perception of liveness. Alongside visual cues like time of day and weather, the organizers of "Our World" sought to convey to viewers the arrival of satellite communications infrastructure by opening the broadcast with live images from around the globe of one of the most singular temporal moments in the human life cycle: the moment of childbirth. In a 1967 memo describing the show's goals, Singer referred to the broadcast's planned opening sequence, a series of consecutive segments from maternity wards on different continents, as "flex[ing] our muscles," an astonishing appropriation of women's labor in childbirth in service of the display of (masculine) technical power.[22]

Yet the selection of one of the most viscerally immediate moments in the human life span did not seem, to the BBC's producers, entirely sufficient for marking the spectacular transcendence of distance accomplished by "Our World." The series of mothers and newborn babies—from Japan, the UK, Mexico, Denmark, and Canada, depicted in the broadcast's opening sequence—were distinguished not only by the announcer's description of where each birth was taking place or the local language used by the doctors and other participants in each segment. The producers also chose to subtly differentiate the women by race, depicting them with more or less concern for the women's privacy and bodily autonomy in ways that sought to convey a racialized hierarchy. While the white mother and baby in Denmark both appeared fully clothed and neat, clearly captured some time after delivery, the Mexican mother was filmed while still giving birth, her spread knees visible from behind her head, her face never shown.[23] In Edmonton, Canada, a Nehiyawak (Cree) First Nations mother and baby, shown last, were presented as the passive beneficiaries of modern medical expertise and explicitly racialized. The Nehiyawak mother is shown receiving her baby from a white nurse, while the Canadian on-air host explains that this is a dedicated hospital exclusively for indigenous women from the Canadian far north, and this mother is seeing her baby for the first time. The host described the new baby girl as having "the jet-black hair of her ancestors" and described life in her indigenous community as dangerously harsh. If the new baby were to survive the "difficult" first year of life in the "rugged north bush country," the host noted, she could expect a life span of "sixty years."[24] This differential presentation of the various mother-baby pairs around the world was not merely incidental to the broadcast's goals—it reflected its organizers' desire to visualize their technical feat using racialized, imperial hierarchies. For a program focused almost exclusively on the developmentally uniform Global North, Singer and his colleagues struggled with how to convey physical distance without referring to the cultural and social distance of empire.

The desire to make traversal of space in "Our World" evident to viewers via an imperial visual language also shaped the BBC's interactions with its Eastern European counterparts throughout the planning process in 1966–1967. In initial proposals to the Soviet side, the BBC suggested that the Soviet inserts come from places specifically associated with Russia's imperial expansion in the eighteenth and nineteenth centuries: namely, Crimea and Central

Asia. However, Molniya Earth stations in Crimea and Central Asia were not yet fully operational by the summer of 1967, putting those locations out of reach.[25] Soviet counterproposals emphasized Soviet contributions to the victory in World War II, a theme that did not interest the BBC organizers. Fokin and Nikolai Mesiatsev proposed that Soviet Central Television provide a segment from the war memorial at Mamaev Kurgan in Volgograd; Singer noted to colleagues that the BBC had rejected this proposal: "This we resisted," he reported, "and it has now fallen out."[26] In lieu of the Volgograd segment, Soviet Central Television negotiators suggested an insert from Sverdlovsk, an idea that Singer and others at the BBC felt was "to put it mildly, a static one."[27] Nonetheless, the Sverdlovsk segment stayed in, as part of a scaled-back plan for Soviet participation with segments from Moscow, Sverdlovsk, and Vladivostok in the Soviet Far East.

BBC organizers also worked to present Eastern European participants as technologically behind Western Europe. An April 18, 1967, memo from the Polish television representative Maryla Wisniewska to Singer asked why the proposed subject matter for their segment had been rejected. "We do not quite understand," Wisniewska wrote, why Polish television could not show workers "who complete their education with an aid of television." Surely, she argued, the need to continually improve "one's professional qualifications at the current speed of development of science and technics" was self-evident.[28] The proposed Polish segment reflected Eastern European public broadcasters' temporal and political claims about television technology, which emphasized the ways that television was contributing to the gradual construction of communism, or at least the ongoing, everyday betterment of society. Both possibilities fit squarely within what Sabina Mihelj and Simon Huxtable identify as two distinctively socialist television temporalities: revolutionary time and socialist time, with the former based on a communist telos "beginning with revolution and ending in a communist future," and the latter referring to a broader conception of temporality, "any temporal practice that is distinctly socialist."[29]

The Poles, in turn, rejected a BBC counterproposal that would have shown Polish passersby "watch[ing] tv sets through shop windows," an image suggesting that, rather than actively using communications technology to advance social goals and popular technical literacy, Polish citizens were largely still excluded from access to modern communications networks.[30] At stake in these two conflicting proposals for the content of the Polish segment of "Our World" were claims about both the temporal nature of the socialist bloc's membership in the modern, satellite-mediated world—were the Poles behind the West in the availability of modern communications technology in the home, leaving Polish citizens to watch "Our World" and other television broadcasts from the street?—and the extent of the Polish state's agency in using television technology to represent a distinctive socialist temporality in which, day by day, socialism was to be built.

The technical coordination of the program was likewise an arena for competition and opposing claims about the division, representation, and control of space. Despite the prevailing rhetoric about the creation of a single infrastructure, the coordination of the broadcast required dividing the globe into spheres of influence behind the scenes. After a July 21, 1966, meeting, Andrew Wiseman and Singer reported explaining to Soviet Central Television's Georgii Ivanov that "we were dividing the world up into three zones," defined, it seems, largely in terms of both existing technical and governing infrastructures (i.e., OIRT and the EBU) and geopolitical influence.[31] The BBC told Soviet Central Television that it would have preeminence within its own "zone" (the OIRT zone, made up of Soviet bloc countries) in the planning process, suggesting that the broadcasting plan would effectively confirm the Soviet Union's political preeminence within its bloc.[32]

In its own internal documents, however, BBC and EBU staff represented Moscow and its Eastern European network quite differently. Two months before the broadcast of "Our World," the EBU's Technical Center circulated a special notice outlining the technical protocol for the coordination of satellite feeds and the organization of work into smaller broadcast zones. In this document, the EBU divided the broadcast's network into four, not three, zones: the West Zone, the EBU Zone, the OIRT 1 Zone, and the OIRT 2 Zone, with control centers in New York, Brussels, Prague, and Moscow. Figure 2.1 depicts the hierarchy of these zones, and, in addition to being subordinated to the "World Switching Centre" and "Master Control" in London, Moscow is placed as the most distant zone from the center.[33]

An accompanying map (figure 2.2) showing the respective control centers further strengthens the visual peripheralization of Moscow within this technical network. London is placed in the very center of the map, competing only with Brussels. New York covers the entire left side of the map, as the node linking North America to Australia and Tokyo. On the right side of the map, Prague, not Moscow, is most visually prominent and serves to link the

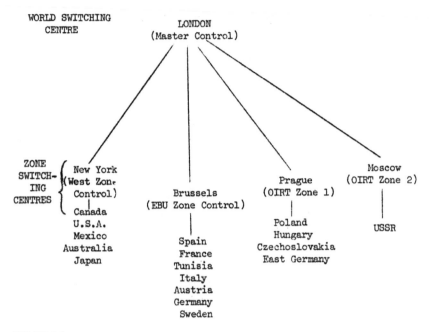

FIGURE 2.1

Organization of "Our World" control and switching centers. Reproduced with permission, BBC Written Archives Centre.

broadcasters of Eastern Europe together. Moscow is situated in the upper-right corner, with relatively few circuits connected to it. The overall impression is that Moscow, as the control center of OIRT Zone 2, makes up only a minor part of the worldwide broadcasting infrastructure created for "Our World."[34]

Moscow's peripheral position in these network maps belied the importance of traversing the Iron Curtain to the original conception of "Our World." They also anticipated the relative ease with which Eastern European participants would be removed from the broadcast after the Soviet withdrawal. Since they were already on the very edges of the control network's map of the world, it was easy to simply cross the socialist countries' network nodes off the map.

An article written by the BBC project leadership for the British magazine *Radio Times* likewise emphasized the way in which this purportedly apolitical infrastructural project was designed to highlight the centrality of London. "All the participating nations have pooled their skill and resources and shared the cost," the article noted, but it continued, "Perhaps we at the BBC may be forgiven for taking a special pride in the fact that the project was

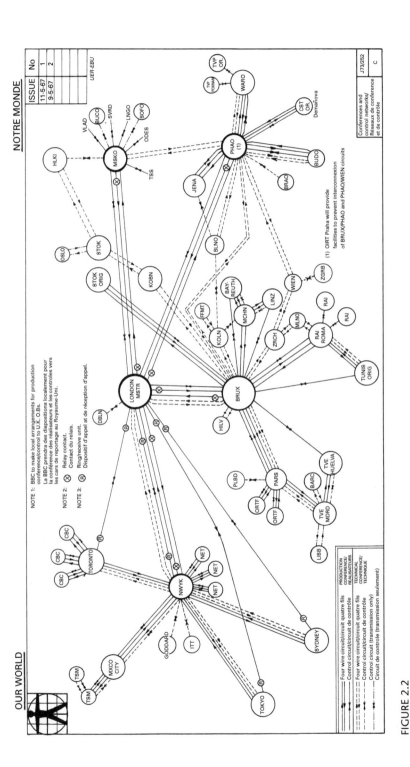

FIGURE 2.2

"Our World" network map. Reproduced with permission, BBC Written Archives Centre.

born here in London, carried to completion under a Project Editor who is a BBC man and that on Sunday night it will be a BBC team in London which will be controlling what is certainly the most complex, and perhaps also the most hopeful, event in television history."[35] Together with the BBC's rhetoric, which celebrated London's position at the center of a new global network, the planning process made clear that, while the BBC understood the importance of presenting the ad hoc infrastructural network of "Our World" as multicentered and confirming the more powerful participants' geopolitical positions, from the BBC's perspective, this broadcast would make London an imperial global communication capital once again.[36] This was even further accentuated in the broadcast itself when the host, Cliff Michelmore, announced: "And here in London, England, is the center of the web, the control room of the whole program. From here it goes out to something like 170 million television sets in 24 countries."[37]

The renewed imperial networks that the BBC imagined extending outward from London, however, were now presented as optimistic rather than exploitative. The broadcast was announced, in a British *Radio Times* article, in epochal terms: "Our World" "gives hope," the article announced, "that perhaps television, in bringing together sovereign states in every corner of the earth may be like the great railways of the 19th Century which linked the scattered, diverse communities of the United States into one great nation."[38] The broadcast was thus presented as an explicitly imperial infrastructural project, one with the potential to endure over time and act upon the political geography of the globe. Like the "golden spike" that finally connected the US rail system, the broadcast would join vast territories under a single technical regime, launching a new era of mutual communication and visibility.[39] Yet the BBC authors of this *Radio Times* article were also at pains, in public rhetoric, to downplay their own central and controlling role in the creation of this live, immediate, and imperial network: immediately after comparing the broadcast to the closing of the US frontier, the authors reassured their readers that "Our World" "is a cooperative venture," constructed on terms of equality, cooperation, and peaceful, nonpolitical objectives.[40]

In fact, however, as the BBC's American railway metaphor hinted, this ostensibly politically neutral, cooperative network was also, potentially, an opportunity for empire building, at least symbolically—a chance for participating powers to represent themselves as central to a vast new media infrastructure. Despite the BBC's triumphal rhetoric about its central place in the broadcast's infrastructure, the success of the Soviet side in negotiating a new broadcast date, June 25, rather than the June 21 summer solstice—a change that in turn meant abandoning references to planetary, cyclical time in the broadcast's name—suggests the significant impact of the BBC's Eastern European negotiating partners in shaping the final broadcast.

It would be easy to see the eventual Soviet withdrawal as a kind of victory for the BBC side, which now could place itself at the center of the broadcast and assert its civilizational and technical superiority unchallenged. In fact, however, the Soviet bloc withdrawal from "Our World" tells a more complicated story, in which the bloc continued to assert its own temporal and spatial priorities in the construction and representation of satellite infrastructure. This brief episode in broadcast history, as well as its use of communications satellites, are reminiscent of the much longer process of global time reform and the introduction of a universal time regime. Negotiations over universal time zones and a world calendar was also, as Vanessa Ogle has showed, an uneven process with "contrasting interpretations of the consequences and meanings of interconnectedness," in which time served as an "intellectual and institutional device for imagining the world as global and interconnected."[41] The competing temporalities of "Our World" had similar consequences, reshaping not only the global network itself but also the way in which the planetary was imagined and understood. Just as Soviet and US networks of satellite tracking stations produced two competing models of the shape of the Earth, the competing temporalities of "Our World," the references to and disputes over planetary and cyclical time, exposed how a single global system could not create a universalized globe but instead acknowledged the unevenness and nuances of the planetary.

WITHDRAWAL

The Soviet decision to withdraw is traditionally portrayed as the result of external, Cold War military-political events. Around lunchtime on June 21, 1967, the chief editor of "Our World," Aubrey Singer, received a telex from Deputy Chairman Georgii Ivanov of the Soviet Union Radio and Television Committee. The telex was brief and explained that the broadcast was supposed to strengthen the mutual understanding and friendship between nations; however, the Six-Day War, which the Soviet telex described as "a plot of certain imperialist forces, primarily the U.S.A., against the Arab peoples," conflicted with this goal. Furthermore, since a number of the countries involved in "Our World" took part in this "slanderous campaign against the Arab countries and the peaceful policy of the Soviet Union and other Socialist states," Soviet Central Television was refusing to participate in the broadcast.[42] Drawing on a longstanding rhetorical strategy on both sides of the Iron Curtain that presented scientific cooperation and cultural exchange as entirely apolitical and disconnected from international political events and relations, the telex framed the withdrawal as a political matter, unrelated to the longstanding cooperative ties and shared commitment to global liveness between Central Television and the BBC.

Yet although the cancelation was politically useful for Soviet–Middle Eastern relations—Soviet Central Television received cables from Middle Eastern leaders thanking them for their gesture of support—the withdrawal from "Our World" may also have been motivated by technical factors that were the product of the Soviet Union's specific ideological and temporal approach to infrastructural projects. In June 1967, most of the network of ground stations for the Molniya satellite system was not yet complete and in service. Reports in the Soviet Ministry of Communication's archives indicate that only two of the planned Orbita Earth stations were fully operational between September 1967 and November 1967, with nineteen more coming into operation in December 1967 and January 1968.[43] This may have been partly due to construction delays; many Orbita Earth stations were located in areas of permafrost, where the instability of the soil could disrupt construction and infuriate bureaucrats.[44] As had been the case with the construction of local television stations, Soviet satellite ground stations were built using a variety of funding sources, with some paid for entirely by local state enterprises, others by the Regional Party Executive Committee (Oblispolkom), and others receiving central funds.[45] These diverse, highly specific local arrangements produced significant delays in some areas, which the Ministry of Communications was obliged to address. But the most important factor shaping the timeline for constructing Orbita Earth stations was the approaching fiftieth anniversary of the October Revolution on November 7, 1967.[46] From the perspective of the Communist Party Central Committee and the State Committee for Television and Radio Broadcasting, which had launched plans for the expansion of Soviet Central Television to all of Soviet territory long before the initiation of planning for "Our World" in 1965, the most important target date for readiness was the revolutionary anniversary in November, not "Our World" in June.

In light of the multiple conflicts and incompatibilities that the planning of "Our World" revealed, the Soviet withdrawal seems overdetermined—far from the "intrusion" of conflict in the Middle East into purportedly apolitical broadcast cooperation. The transnational infrastructural project that produced "Our World" was shot through with conflict and incommensurability between the sides, often expressed temporally and spatially. However, the shared attraction of televisual liveness and global presence on both sides must be taken seriously. These shared ambitions helped facilitate the construction of multiple transnational networks centered on different metropoles.

FORGETTING "ONE HOUR IN THE LIFE OF THE MOTHERLAND"

Yuri Fokin's presentation of "One Hour in the Life of the Motherland" as the original idea of young Soviet journalists in August 1967 concealed the ways in which "One Hour in the Life of the Motherland" rehashed and expanded what had been programming and technical plans for the Soviet Union's participation in "Our World," as well as borrowing additional elements from "Our World," such as broadcasting from a maternity hospital. Moreover, "One Hour in the Life of the Motherland" reframed the task of celebrating satellite broadcasting in direct response to the major points of conflict between Soviet, Polish, and Czechoslovak participants in "Our World" and that broadcast's BBC organizers. If "Our World" had showcased London as the new imperial metropole, the controlling and directing center of new Space Age satellite communications infrastructures stretching around the globe, "One Hour in the Life of the Motherland" positioned Moscow as the renewed imperial center from which the show's elaborate network of space and Earth connections was imagined and coordinated. "One Hour in the Life of the Motherland" also responded directly to the efforts of the "Our World" broadcast to depict the socialist world as having less access to modern technology and scientific achievements, highlighting Soviet medical achievements and featuring schoolchildren in symbolically "underdeveloped," ethnically non-Russian Soviet Tajikistan. Where the BBC had proposed showing Russian schoolchildren in Leningrad—suggesting the tutelary position of even the most advanced regions of the socialist world— "One Hour in the Life of the Motherland" reversed these messages, putting European Russia at the center and its own internal, racialized periphery in the position of tutelage.

The reversal in "One Hour in the Life of the Motherland" of the imperial hierarchies proposed by BBC organizers for "Our World" suggests the
significant overlap between the BBC's and Soviet Central Television's visions
for these broadcasts: both sides sought to renew and reaffirm imperial hierarchies of technical and cultural development using this new technology. But
"One Hour in the Life of the Motherland" was also shaped by the specific,
revolutionary temporality that had characterized Soviet documentary media
since at least the 1920s. Lisa Parks described "Our World" as layering and
juxtaposing various temporal structures—population growth, ticking metro-
nyms, the planetary rhythms of sunrise and solstice, economic modernization, broadcast duration, and so on—to establish "new forms of planetary
management and control."[47] The one form of temporality that "Our World"
could not assimilate, however, was the one that animated "One Hour in the
Life of the Motherland," in which world events, selected and assembled by a
skillful editor, would convey to viewers the progress toward the construction
of communism and the arrival of world peace after the successful revolutionary synchronization of the world—generating a mood of elation and joy.

An internal discussion of "One Hour in the Life of the Motherland" from
early October 1967 give a sense of how that broadcast's producers understood
its connection to the revolutionary anniversary and envisioned its affec-
tive impact on viewers. "Life in our country on the first day of the holiday
month . . . at the moment of fulfillment of important and interesting events,
will appear before the viewer," one of the producers reported. "'One Hour
in the Life of the Motherland' will convey a sensation of our Motherland's
immensity," he continued, "of the elevated atmosphere in various spheres of
life in the Soviet state . . . of the breath of time, the clear rhythm of life."[48]
Unlike "Our World," with its grim, eugenicist portrait of the present as a race
between Western scientific innovation and uncontrolled reproduction in the
Global South, "One Hour in the Life of the Motherland" depicted a euphoric
revolutionary synchronization of Soviet space.

Yet this elevated rhetoric about "One Hour in the Life of the Motherland,"
emerging, notably, in advance of the actual broadcast, also helps us understand another key difference: the fact that this broadcast has been almost
totally forgotten, even within the quite extensive memoir literature by Soviet
Central Television executives during this period. The script is not preserved
in archives and there is no recording of the show, and thus the only sources
marking the existence of this quite high-profile broadcast are a handful of

mentions in Soviet Central Television's archives and several articles in the professional press that reviewed and analyzed the broadcast in the months that followed.[49]

What can explain this significant differential in commemoration between "Our World" and "One Hour in the Life of the Motherland"? One significant factor may be that satellite television broadcasting became infrastructure— that is, it became routine—more rapidly in the Soviet Union, where the Orbita satellite system delivered television daily to most of the eastern and far northern parts of the country from November 1967 onward; satellite television remained associated with exceptional, high-profile, global media events for much longer outside the Soviet Union, where it was not in daily use. Moreover, a new flagship Soviet TV evening news program called "Programma Vremia" [Time], launched just two months later on January 1, 1968, featured a format nearly identical to "One Hour in the Life of the Motherland." Like "One Hour in the Life of the Motherland," "Programma Vremia" featured live and recorded inserts from around the Soviet Union and the world and was promoted via almost identical rhetoric of panoramic breadth and the clear rhythm of time.[50] The program also centered satellite ground infrastructure and a whole-Earth image in its famous sign-on.

The most compelling argument, however, returns us to the ideological and temporal premises of "One Hour in the Life of the Motherland," which distinguished it from "Our World," even though these two satellite spectaculars were very much the product of interaction and mutual influence. Instead, "One Hour in the Life of the Motherland" proposed that communications satellites would offer simply a more-live, more-instantaneous version of what already had been a central genre of Soviet documentary media since the 1920s. Then, nearly identical ideas of live presence and immediacy, global and all-Union synchronization, as well as the use of film, radio, and later television to artistically and politically unify a vast, ethnically diverse world, had inspired Soviet filmmakers like Dziga Vertov to make films like *A Sixth Part of the World, Lullaby,* and others, which offered similar promises about the revolutionary impact of live immediacy and virtual travel.[51] Similar ideas motivated both Stalinist and post-Stalinist journalists, including Maxim Gorky and Aleksei Adzhubei, each of whom oversaw photographic and journalistic projects entitled "One Day in the World" that brought together, in a single published volume, the events of September 27, 1935, and September 27, 1960, respectively.[52] Like "Our World," and, indeed "One Hour in the Life of the Motherland," Adzhubei's 1961 "One Day in the World" book also employed childbirth as a motif, opening with a global chronology that began with the birth, at midnight, of a baby boy to Tatiana Pakhomova, a nurse in Moscow.[53]

By the late 1960s, a revival of Vertov's ideas in particular was in full swing with the publication of a new edition of his collected writings and the identification, by enthusiastic young television producers and editors, of their new medium with Vertov's ideas about the power of live presence (the *effekt prisutsvie*). Yet television workers and Communist Party ideologists, by the late 1960s, faced a growing realization that liveness alone was not inherently exciting, nor was it capable of ensuring that viewers drew the right conclusions from the documentary images of Soviet life before them. As Gennadii Sorokin, an instructor in the Central Committee, wrote in a review of "One Hour in the Life of the Motherland" published in the professional journal of the Union of Journalists, "Even a non-specialist understands that technology on its own does not create any kind of 'effect of presence' [*effekt prisutstviia*] or sense of the immediacy [*siiusekundnosti*] of the events happening on screen."[54] The show's contents, Sorokin pointed out, hardly seemed worth the enormous effort required to transmit them live via satellite. "Was it worth putting in motion this gigantic communications system," Sorokin questioned, combining "mobile TV stations, radio relay lines, and communications satellites, in order to show on Moscow screens how gears turn in Cheliabinsk or how a reporter from Tashkent television talks his way through a boring script?" He pointed out that the inability to fulfill the promises of broadcast technology was a larger problem in the Soviet media system that long predated satellite communications. It happened all over, he noted, that radio relay and cables lines were used to broadcast old movies that most provincial cities already owned.[55]

Sorokin did single out two moments in "One Hour in the Life of the Motherland" that, he said, "genuinely created a wonderful, strong impression of our inclusion in the life of the country": the fact that the broadcast was timed to take place between two changes of the guard at Lenin's Mausoleum on Red Square, and an announcement that an airplane, whose crew would appear later in the broadcast, was still in the air.[56] Yet even these two moments in "One Hour in the Life of the Motherland" were not original to that broadcast or even dependent on satellite communications: these ways of making liveness visible were well established on Soviet television and had been included in previous Soviet Central Television broadcasts, most of which were live (though not via satellite) by default, given the shortage of film stock and slow integration of video. Moreover, even if we accept that these unoriginal strategies for signaling live immediacy to viewers genuinely impressed, Sorokin pointed out that even these successful moments "only underlined the ineffectiveness of other moments" in "One Hour in the Life of the Motherland." The handful of other reviews that were printed in the professional press in the months after "One Hour in the Life of the Motherland" drew similar conclusions and likewise stressed the problem of the gap between the show's exciting technical firsts and the banal content.

The live satellite linked inserts from cities around the Soviet Union not only failed to live up to the elaborate infrastructures linked to transmit them, they were also lost within a sea of similar documentary content that local television stations in particular spent most of their time producing and exchanging. On a trip to Tbilisi's television archives to search for a script for "One Hour in the Life of the Motherland," we did not find it. But we did observe that in October and November 1967, Tbilisi's television station produced an enormous number of other recorded documentary features, for broadcast or exchange with other Soviet- and socialist-bloc television stations, that presented local economic and cultural life in ways that were essentially indistinguishable from the "One Hour in the Life of the Motherland" inserts.[57] No wonder this particular script—so technically original but textually banal—was neither preserved nor remembered.

The story of "One Hour in the Life of the Motherland" thus reflects the fundamental tensions underlying efforts to promote and celebrate the promise of communications satellites: despite their Space Age technology, communications satellites offered a world public only expanded access to existing earthly media flows, not a break with Earth and the human past. This tension was especially evident in the Soviet Union, moreover, since, on the one hand, the fantasy of electronic media flows transcending geopolitical borders was especially intense for some Soviet viewers and especially unwanted by Soviet officials, and, on the other, documentary film and journalism enabling virtual travel around the Soviet landscape had been done to death.

While "Our World" has been somewhat better remembered, its commemoration and afterlife on YouTube is chiefly linked to a live performance on the show by the Beatles. Unlike other space technology, the promise of communications satellites was limited to the media that they could distribute. Thus, 1967's satellite "spectaculars" were rather unspectacular: communications satellites could offer only intensified access to the sights and sounds of other humans elsewhere on our planet, not transcendent experiences of a world beyond Earth. Whether the future that these two satellite spectaculars promised was ominous and uncertain, as "Our World" proposed, or euphoric and revolutionary, as "One Hour in the Life of the Motherland" suggested, the effort to visualize and present to a global public the promise of this new space infrastructure was limited to tropes of modernization, development, and pleasurable media consumption that were not original to satellite communications.

CONCLUSION

The Soviet broadcast did not go unnoticed by Western broadcasters. The response to "One Hour in the Life of the Motherland" by BBC producers and others involved in "Our World" reiterated the claim, during production of the latter, that the Soviet Union and its allies were hopelessly behind in developing and using the modern space technology that both broadcasts showcased. In a letter dated November 7, 1967, Noble Wilson at the BBC told Peter Pockely at the Australian Broadcasting Commission about the broadcast of "One Hour in the Life of the Motherland," noting the significant similarities between the broadcasts, particularly segments from a maternity ward and other ideas developed during the planning of "Our World" (but not, in fact, original to that broadcast). Hinting at the real technical rivalry underlying the show's concept, Wilson made derisive comments regarding the sound and picture quality of the show, commenting that "a sequence from a woollen [sic] mill in Tashkent had all the quality of an early Daguerreotype." He continued with an upbeat, but still profoundly condescending, observation: "Watching the program at this end, what I think we learned was that . . . Soviet Television was technically capable of mounting an exercise of this nature and that they could run it on time," which, he concluded, was "heartening for those of us who hope that one day they really will join in and sing."[58]

Wilson was referring to concerns throughout the planning of "Our World" about whether Soviet Central Television would be able to provide the necessary satellite links to bind the vast country together, and ultimately link it to the Western sphere of broadcasters. These remarks continued to position London as the center of an apolitical technical modernity that the Soviet state was ostensibly still struggling to join; the Soviet Union's ability to run a satellite broadcast "on time" suggested, in his mind, a step toward a harmonious future, figured, in his metaphor, as singing. Again, he presented the Soviets as backward, recalcitrant, and not yet full participants in this modern choir. Remarkably—and much like Fokin's invented origin story for "One Hour in the Life of the Motherland"—Wilson's comments effaced the long history of Soviet–British live broadcasting cooperation, including the cooperation of European networks in the live broadcast of Moscow's May Day parade and Yuri Gagarin's return from space. Unlike "Our World," those broadcasts made Moscow the center of modern live communications infrastructures. Perhaps this was why they were so easy to forget, from Wilson's perspective, just as Fokin found it easy to describe the Moscow-centric "One Hour in the Life of the Motherland" without referring to either the BBC or "Our World."

Yet Wilson's claim was far from uncontested. The Soviet "One Hour in the Life of the Motherland" broadcast, whatever he thought of it, offered its own, competing temporal and spatial mapping of global satellite modernity for its national audience—one that unfolded within Soviet national boundaries, but which nonetheless closely resembled "Our World."

The idea of "global presence" that Parks characterizes as a "Western fantasy" and links to Western discourses of modernization, was thus very much ideologically contested terrain during the years in which "Our World" was planned and produced. Soviet cultural and political elites had long presented Moscow, in a variety of contexts, as the center of an alternative modernity—one that was equally technologically advanced, but linked to progressive ideals and a millenarian account of historical time.[59] The ideal of global, live satellite broadcasting proved quite flexible and open to reuse by the Soviet side for imagining a Soviet-led satellite network—the one that was eventually successfully institutionalized as Intersputnik.

When we include the interactions between the BBC and Soviet Central Television in the story of "Our World," the broadcast looks somewhat less successful than the BBC claimed, both internally and in public. Rather than unproblematically affirming the networked superiority of London and other global capitals, the claims underlying the "Our World" broadcast remained open to contestation from Moscow and other socialist world participants both before and after it was produced. The transnational broadcast's claims to liveness and global presence were both facilitated and undermined by a complex set of personal relationships, rival rhetorical claims, and material infrastructures. The technical, spatial, and symbolic conflicts that shaped the broadcast—the underlying geographies of liveness at work, we argue, in any transnational live broadcast event—profoundly shaped the "Our World" broadcast and its Soviet counterpart, "One Hour in the Life of the Motherland." The challenges of creating a global satellite infrastructure problematized the ability of "Our World" to serve as a triumphant display of Western technical superiority, based on the claim to universal liveness and global presence. Instead, they reveal a more complicated, fragmented picture of the broadcast and reception of "Our World," in which traces of conflict and the active role of unequal participants are made visible.

3 FRAGMENTED FROM THE BEGINNING: THE ENTANGLED ORIGINS OF INTELSAT AND INTERSPUTNIK

On December 17, 1973, Dmitrii Ustinov, the chair of the Military Industrial Commission of the USSR Soviet of Ministers—in lay terms, the head of the Soviet military-industrial complex—received a report on the status of US aerospace research and development from his former subordinate, Georgii Nikolaevich Pashkov, now retired and working as a consultant. Pashkov focused mostly on one strategic area of international aerospace rivalry: satellite communications. Summing up the differences between the Soviets' non-geosynchronous Molniya-series communications satellites and the US's geosynchronous INTELSAT-IV series—named and built for Intelsat, the US-led global satellite communications organization created in 1964, Pashkov concluded, simply, that "their satellites are much better than ours."[1]

Pashkov's admission that Soviet communications satellites were, as he saw it, technically inferior—they had far fewer channels that could carry telephone or television transmissions, for one—seems to confirm a highly recognizable Cold War framework that still dominates the limited Western historiography of satellite communications, just as it shaped the BBC's response to the Soviet withdrawal from the "Our World" project. Even as technology beyond the nuclear arms and space races has come to enjoy greater attention in general histories of the Cold War, the history of Cold War technology continues to be told almost exclusively as a story of Soviet failure and US triumph.[2] Insofar as they address Soviet participation in the first two decades of satellite communications at all, the handful of existing histories of satellite communications have largely kept this competitive, binary focus. James Schwoch, in his rich and insightful history of the role of satellites in transforming television into a global medium, concludes by arguing that the development of communications satellites marked a larger victory for American science—one that paved the way for US domination of the process of economic and technological globalization.[3] Fittingly, Schwoch ends his story in 1969, with Intelsat carrying images throughout the globe of Neil Armstrong's space walk, integrating the history of satellite communications with the history of the Space Race by giving these two stories a shared ending: a US scientific victory, anointed by an awed global television audience. Together with Lisa Parks's foundational work on satellite cultures and infrastructures, Schwoch has helped to found an emerging field of the critical study of satellite media infrastructure—a field that is, nonetheless, implicitly predicated on the assumption that US state and corporate power require our critical attention precisely because they were victorious in the Cold War technology race.[4]

It is striking, then, that North American telecommunications officials and scholars writing in the early 1970s, at roughly the same time as Pashkov's admission of technical defeat, almost universally agreed that the outcome of Intelsat's renegotiation of its governing regulations was a resounding *failure*.[5] These commentators focused not on the relative superiority of US satellite technology, but rather on the stated objective of US policy with regard to Intelsat: the creation of a single, global network, within which the US retained the dominant role.[6] Although contemporaries noted that disagreements within Intelsat itself were legion, they emphasized that the greatest failure was the inability to negotiate Soviet entry into Intelsat, which would "remain one of the real pities of the development, to date, of satellite communications."[7] The outcome of the Intelsat negotiations of 1969–1971, in other words, was not an unequivocal US victory but rather the collapse of one American space policy objective of the 1960s: that of a single, global satellite communications network under US leadership. In the coming years, many more regional satellite communication consortia would be formed, laying the foundation for today's multicentered communications satellite industry—an outcome that should not surprise us, given that previous single-system dreams for new media infrastructures had also proved impossible to realize.

As these conflicting accounts suggest, a binary framework of US–Soviet triumph versus defeat, as well as a focus on technical achievements in isolation from the human institutions and material infrastructures needed to support them, cannot accommodate the history of communications satellite infrastructure. First, while there was plenty of rivalry and competition, it was as often between the US and its Western European allies as it was between the US and Soviet Union. Moreover, this rivalry was commercial as much as geopolitical, in part thanks to the key role of communications satellites as evidence that both Cold War superpowers' massive investments in the Space Race could generate revenue and benefit regular citizens.[8] Indeed, the creation of and evolution of the US-led Intelsat network and its Soviet counterpart and apparent rival, Intersputnik, in the 1960s and 1970s, reveal a great deal of political and economic common ground in Soviet, European, and American understandings of what form satellite communications infrastructures should take and how they should be used.[9]

Unlike the autarkic, secretive Space Race projects of the 1960s, therefore, the development of Intelsat and Intersputnik was forged in extensive behind-the-scenes interactions, leading to what became the functional integration of the two networks by the early 1970s. In fact, in September 1969, before Intersputnik was officially operative, Intelsat officials already had noted during a conference with participants from ten countries, including the Soviet Union and the US, that an integrated system was fully possible. "Technical compatibility between the two major satellite communications systems—Intelsat and Intersputnik (through the Soviet Orbita)—is not difficult to obtain," the Intelsat officials noted. "Their orbital systems are complementary; their frequency plans can be coordinated with a single ground station being able to operate in either system: and their transmitting and receiving equipment can be adapted for operation in both systems."[10]

Thus, while Western scholars have long assumed that Intelsat and Intersputnik were entirely separate intergovernmental membership organizations, installing the Iron Curtain into the skies, archival records from both sides reflect substantial interaction (among the US, the Soviet Union, and Eastern and Western European countries, as well as with countries in the Global South), mutual influence, and overwhelming agreement that satellite communications infrastructure built in the 1960s and 1970s should support an integrated, global, and commercial satellite communications system. Despite this common ground and mutual commitment to global integration, however, the eventual creation of Intersputnik as a competitor to Intelsat in 1971, with the tacit support of several European governments, also had a significant impact on Intelsat's own organizational structure and laid the groundwork for the eventual regionalization and fragmentation of satellite communications infrastructures around the globe, frustrating US visions of a global, US-led monopoly under Intelsat. The story of the creation of two—ostensibly separate and rival, but not entirely either—satellite communications networks thus fits neatly within a set of new Cold War histories of technology and scientific exchange that, rather than refighting the Cold War by comparing technical achievements on both sides, emphasize the circulation of people and ideas across borders and seek to identify the ways in which Cold War geopolitics contributed to, rather than impeded, infrastructural and financial globalization.[11]

What can this more integrated view of the Cold War bring to bear on theories of how infrastructures change over time?[12] The mutual evolution of Intelsat and Intersputnik offers us a case study in infrastructural globalism, in which the world is "produced and maintained—as both object of knowledge and unified arena of human action—through global infrastructures."[13] Large-scale historical narratives of information infrastructures have traced both the emergence and institutionalization/infrastructuralization of this conceptual globalism and the eventual fragmentation or splintering of global infrastructures under the pressures of privatization and inequality. Stephen Graham and Simon Marvin, for example, have described the evolution of infrastructural networks and cities via a "parallel set of processes . . . in which infrastructure networks are being 'unbundled' in ways that help sustain the fragmentation of the social and material fabric of cities."[14]

The idea of infrastructural fragmentation as a process tends to construct a past in which such splintering was *not* evident. Yet the metanarrative of the rise and fall of truly global infrastructure appears far more complicated up close. For the case of satellite communications, we argue that there was no such pure moment of wholeness before splintering began—indeed, such moments of wholeness arguably can never exist except in the realm of imagination and rhetoric. Rather, the geopolitical and commercial interests that shaped the formation of satellite communications institutions and infrastructures did so in ways that baked future fragmentation into those institutions and infrastructures from the beginning.

At the same time, we emphasize that splintering took place primarily at the level of political and economic control—the failure of US plans for a global monopoly under Intelsat did not preclude the eventual technical and commercial integration of satellite networks across the Iron Curtain during the 1970s. Indeed, the outcome of the crucial early years (1967–1971) was a global network that *was* functionally integrated, but in which political control and profit were more decentralized than US officials had hoped. Yet here too, a Cold War framework of victory and defeat does little to explain the origins of contemporary satellite infrastructures or the impact of these key negotiations. The commercial, political, and scientific cooperation and interaction in the realm of satellite communications before 1991 both laid the groundwork for international cooperation in human space exploration (as we demonstrate in chapter 5) and facilitated the very rapid commercial integration of Soviet aerospace infrastructure into the global space economy after 1991.

GLOBAL PROMISES AND THEIR CRITICS

American rhetoric about the transformative potential of global satellite communications gave little indication of the extensive resistance US plans faced from the start, clothing the dream of global communications in the well-documented rhetoric of infrastructural utopianism.[15] President John F. Kennedy's July 24, 1961, remarks, announcing plans to form a global satellite communications network, presented participation by "all nations" in such a network as "in the interest of world peace and closer brotherhood among peoples throughout the world."[16] Such sweeping and vague generalities, of course, could not paper over the significant conflicts of interest inherent in negotiating the terms of a new, binding political and commercial relationship with numerous other countries. As Hugh R. Slotten and others have demonstrated, Intelsat's organizational structure was shaped by conflict between competing interests within the US government, especially with regard to the decision to create a private corporation, COMSAT, to be the US representative within a future Intelsat network. Despite its status as a private, for-profit corporation, COMSAT was also subject to control and oversight by Congress, creating a fundamental tension between its roles as an instrument of US foreign policy and as a commercial entity. Competing visions of Intelsat's structure were likewise divided between many separate, bilateral, commercial leasing arrangements, modeled on the undersea cable system, and a multilateral, intergovernmental organization, favored by the Europeans and the US State Department, in which political and commercial decisions would be made jointly.[17] A negotiated settlement, in which COMSAT provided most of the capital and would serve as the manager of a consortium of national telecommunications agencies, was signed in August 1964 with a membership of chiefly European states, whose transatlantic traffic with the US offered the greatest commercial potential. Nonetheless, the dissatisfaction of COMSAT's European counterparts meant that the 1964 accords were designated as just interim arrangements, covering only five years and requiring renegotiation in 1969.

Intelsat's first five years, from 1964 to 1969, were thus riven by significant disagreements among all sides. In the US, COMSAT and multiple agencies within the US government continued to have contradictory understandings of the nature of US objectives within Intelsat. Intelsat's ability to generate lucrative contracts for American aerospace firms made it an important example of how American investment in space produced economic benefits for US taxpayers, especially as the US was drawn more deeply into the Vietnam War.[18] Conflicts between COMSAT and Intelsat's Western European partners continued. The former sought to maximize profits by buying the lowest-cost technology, generally produced by US manufacturers. European member-states, however, wanted a guaranteed share of Intelsat's contracts to develop their own high-tech manufacturing sectors in return for their capital investment in Intelsat.[19]

At the most basic level, Kennedy and Johnson's vision of a single, global network, dominated, under the 1964 Intelsat interim agreements, by the US, produced a great deal of resistance everywhere but in the US itself. From the US perspective, competition from a rival global satellite communications network, or even a limited regional network, was seen as anathema to a successful (i.e., profitable) system. Yet the process of building global governance institutions and infrastructure constantly revealed significant opposition to US economic and technological hegemony from both Western Europe and throughout the Global South.

One of the earliest centers of this resistance was France, which hoped here, as in other spheres of international politics, to counterbalance the global expansion of US power and to retain cultural and economic hegemony in its former colonies.[20] One avenue for this resistance was the pursuit of scientific and technical cooperation with the Soviet Union, which had begun in 1965 to publicly discuss the possibility of an international, Soviet-led satellite communications network.[21] The second half of the 1960s saw growing concerns among US diplomats about the potential for Franco–Soviet cooperation in global communications satellites. These fears were not entirely implausible.

Franco–Soviet experiments in satellite television broadcasting began in 1965 and continued through the early 1970s, featuring joint experiments in color and satellite television broadcasting, the exchange of personnel, and regular meetings in both countries.[22]

French–Soviet joint experiments remained experiments; French diplomats assured their US counterparts that they prioritized their scientific and commercial cooperation with the US and Intelsat over Soviet cooperation. More threatening, however, was the regionalist, multipolar vision of the future of satellite networks that French officials promoted and pursued consistently throughout the 1960s. In 1965, Rene Sueur, the chief engineer for the French telecom agency CNET, told US diplomats that the future of satellite infrastructure could be much like the early years of New York City's subway system, which was initially owned and operated by multiple private companies. Just as it used to be possible to transfer directly from the Interborough Rapid Transit Company (IRT) lines to those of the Brooklyn-Manhattan Rapid Transit Corporation (BRT), the US attaché explained, "Sueur said he visualizes the creation of two partially (if not completely) overlapping world telecommunications systems, COMSAT and a Soviet system, and said he feels it may be necessary to arrange some means for program or message transfers between the two." The attaché's report continued, "The Soviet system [that Sueur] visualizes" would have "a special appeal in Africa and Asia." Sueur observed that countries such as India "might wish to join both the COMSAT and the Soviet systems." Sueur had asked the attaché, "Could not France . . . in some way assist in bridging the gap between these systems?"[23] His transit metaphor cleverly recast the relationship between potential rival satellite networks as regionalized and cooperative rather than exclusively competitive and zero-sum. It also artfully suggested that this division into two separate networks could be temporary since the several transit firms that he referred to, which had each administered parts of the New York City subway system, had been unified in a takeover by New York City decades earlier.

Of course, French officials seeking to retain cultural influence over Francophone former colonies were far from the only voices expressing concern about the political and cultural impact of a single, US-led global satellite network. Intelsat's expansion, accompanied by utopian US rhetoric about peace and brotherhood, took place in a period of significant anxiety about the border-eroding potential of communications satellites and the imbalance of television flows. While the necessity of using specialized Earth stations with very large antennas to send and receive signals from satellites in the 1960s offered states significant control over broadcast content within their borders, this state of affairs was not expected to last long. In the imminent future of direct-to-home satellite broadcasting, commenters worried that a country that disregarded International Telecommunication Union (ITU) frequency allocations, for example, "could make propaganda broadcasts to anyone with a receiver."[24] Such a country might even, this fearful scenario suggested, give away free television sets in targeted regions.[25] Participants on a panel on communications satellites organized by the American Institute of Aeronautics outlined the potential dangers of a single, global commercial satellite network from the perspective of developing nations, and proposed solutions, most of which entailed, at a minimum, "bilateral or regional agreements" that would allow a country to "choose what it wanted to receive."[26] Geoffrey Pardoe, a British aerospace engineer, proposed a more radical approach to prevent uncontrolled direct broadcasts by either superpower—the breakup of satellite infrastructure itself from a single global network to multiple regional ones. "Unless regional satellite networks are developed," Pardoe warned, "you may have 47 frustrated countries unhappy with 'the mighty American machine'."[27] The article closed by noting both the Western European and Soviet plans for their own regional networks, distinct from Intelsat.[28]

It was in this context of hostility to the possibility of US dominance within a single global Intelsat network that President Lyndon B. Johnson renewed US overtures to the Soviet bloc. On August 14, 1967, Johnson used the occasion of his speech to Congress on communications policy to again publicly invite the Soviet-bloc countries to join Intelsat in a single, global communications network.[29] Johnson insisted that Intelsat's mission was apolitical, drawing on well-established rhetoric describing scientific and media exchanges, among other activities, as entirely separate from political goals.[30] "INTELSAT is not a political organization," Johnson insisted, claiming that "it holds no ideological goal except that it is good for nations to communicate efficiently with one another."[31] Nonetheless, Johnson acknowledged the political difficulty of negotiating Soviet entry into a global system in which the US and the COMSAT corporation would play such an important and dominating role.

From the US perspective, gaining Soviet entry into Intelsat was desirable for a variety of reasons. First, the failure to include the Soviet bloc was a significant obstacle to Intelsat's claim to be a truly global network. Many

Western European members, whose support COMSAT needed, also felt strongly that Intelsat should reach out to the Soviet Union and its bloc as actively as possible. Furthermore, as Johnson pointed out, there was "no insurmountable technical obstacle to an eventual linking of the Soviet MOL-NIYA system with the INTELSAT system," and many reasons to do so.[32] It was also politically important for the US to avoid the impression of excluding socialist countries, thereby alienating potential members within the developing world.

The most important reason the US sought Soviet membership in Intelsat, however, was not political but commercial. Internally, both the US and some Western European Intelsat members indicated that the most important reason for including the Soviet Union was to avoid, as one October 1968 memo put it, "unnecessary competition" for Intelsat.[33] Integrating the Soviet Union and its allies would prevent a separate, Soviet-led global network from undercutting Intelsat's prices and allow members to retain control over the lucrative work of manufacturing high-tech components and constructing Earth stations around the globe. By the fall of 1968, however, the Soviet Union was actively engaged in planning for the creation of just such a rival international network.

While the decision to move forward with a Soviet communications satellite network did represent a failure for US foreign policy, the story of Intersputnik's creation also does not map neatly onto the narrative of Cold War high politics. Instead, it reflects the circulation of ideas, people, and influences across Cold War divides, as well as striking commonalities in how US, Soviet, and European elites understood their own interests and the purpose of satellite communications globally.

THE REJECTION OF INTEGRATION? THE INTERSPUTNIK PROPOSAL

Efforts to create a Soviet-led international satellite network that would provide an alternative to Intelsat had begun in earnest by April 1967, at a meeting in Moscow of Interkosmos, the Soviet organization devoted to promoting and organizing international scientific cooperation in space and led by the Soviet Academy of Sciences.[34] In August, the Ministry of Communications presented a draft plan for an international satellite network to an Interkosmos working group. Finally, just after the Molniya network began regular domestic television broadcasting service in the Soviet Union in December 1967, technical specialists from all the future founding members of what would become the Interputnik network (Bulgaria, Hungary, East Germany, Cuba, Mongolia, Poland, Romania, and Czechoslovakia) were invited to Moscow to work out technical plans for such an organization.[35] The working group's correspondence in spring 1968 was characterized by some urgency, since the head of Interkosmos, the academician B. N. Petrov, hoped to officially announce the creation of the organization at the United Nations (UN) meeting on Peaceful Uses of Outer Space in Vienna in August 1968.[36] At a June 1968 meeting in Budapest, the participants approved the proposed articles as working drafts and selected the name "Interputnik," combining the words "international" and "satellite."[37]

The Interputnik draft articles of agreement emphasized that this network was to be founded on very different principles from Intelsat. "This project," the articles claimed "is built on principles of international cooperation, equality, and mutual benefit of all participants."[38] The main basis for this claim was the proposed decision-making body, a council, not unlike the UN General Assembly, in which each member-country would receive one vote, regardless of its level of investment in or use of the network's infrastructure or services.[39] Intelsat, by contrast, was governed by a body that used weighted voting, giving countries that invested in and used the network a greater share of decision-making power, and therefore leaving the US with more than 50 percent of the votes. Like Intelsat, however, Interputnik was intended to be global in its ambition and membership; the articles' authors proposed sounding out "other countries, like France, the Arab countries, India, Pakistan, Burma, and others, to clarify the possibility of their participation in the proposed system."[40] The idea of an alternative, socialist-led international communications satellite network was thus presented as both a rebuke and an alternative to Intelsat.

Yet how strong of a rebuke was it? The Soviet bloc was proposing a network that was explicitly designed to attract nonsocialist members, and in fact strongly resembled what Western European members of Intelsat were seeking in that organization's permanent arrangement negotiations, with the aim of balancing the overwhelming US dominance.[41] The Interputnik proposal also resembled the multilateral structure for Intelsat proposed by both Western European governments *and* the US State Department in the initial negotiations of 1963–1964.

These resemblances between European positions in the Intelsat negotiations and the Intersputnik draft proposal were not coincidental. The Soviet announcement in August 1968 had been supported behind the scenes by France and Switzerland, who hoped, at a minimum, to weaken US influence in Intelsat's upcoming negotiations. At best, they hoped to bring about a proliferation of regional communications satellite organizations, of which one would be Western European. In the months after the announcement in Vienna, they continued these efforts. On September 25, 1968, for example, a diplomat in the Soviet embassy in Washington, V.A. Racheev, met for lunch with his Swiss counterpart in charge of space affairs, one Mr. Steiner. Steiner urged the Soviet Union to release information about Intersputnik's capacity and the date when it would come into service, in order to "strike another blow" to the US position within Intelsat.[42] In response, Racheev pointed out that the timing of Intersputnik's realization was unclear because they did not yet know whether the Europeans, who could help fund it, would in fact join. At the same time, he pointed out, "we are getting the impression that some members of 'Intelsat' would like to speed up the creation of our system only in order to strengthen their position in negotiations with the US, and do not seriously intend to participate in Intersputnik." The best outcome, Steiner insisted, was multiple regional and commercial satellite systems, including US-, Soviet-, and European-led networks.[43] On several occasions, the French made similar overtures to the Soviets on behalf of a European satellite program.[44]

As the regionalist vision articulated by Western European countries in negotiations with the Soviets suggests, however, the idea that Intersputnik was genuinely different in principle from Intelsat was not very well founded. Instead, it represented a well-established negotiating position within Intelsat's own organizational structure. As both sides acknowledged internally, Intersputnik's organizational structure was closely modeled on that of Intelsat. Intersputnik was conceived from the beginning as an independent, commercial entity that would eventually own its own space segment (satellites). As with Intelsat, Earth stations would belong to the countries in which they were located.[45] Soviet claims about the superior egalitarianism of Intersputnik's one-country, one-vote governance structure concealed the fact that the rest of its articles resembled Intelsat's 1964 structure, with only minor adjustments that reflected changes that Western Europeans sought within Intelsat.[46] Thus, despite US arguments at the time and subsequently that the Soviet Union must object on ideological grounds to Intelsat's private, commercial structure, the Intersputnik proposal that was made public in August 1968 reflected not a uniquely socialist alternative to Intelsat, but rather a consensus position forged in interactions across the Cold War divide.

Moreover, via talks with the US embassy and other channels between August and December 1968, Soviet diplomats conveyed their willingness to concede or limit even the only real distinguishing feature of the Intersputnik proposal: the one-country/one-vote structure. In an August 17, 1968, telegram, just days after the Intersputnik proposal was announced in Vienna, the US deputy chief of mission in Moscow, Emory Swank, reported that Soviet diplomats had raised the subject of terms for Soviet entry into Intelsat. Soviet diplomats told their US interlocutors that they were flexible about requiring a one-country/one-vote decision-making body within Intelsat as a precondition to joining, since they knew that was unacceptable to the US. Instead, they stressed that "some assurances re purchase and use [of] Soviet communications equipment in third countries" might serve as an adequate incentive for Soviet membership.[47] In other words, despite retrospective assumptions among Western scholars that the Soviet Union would have insurmountable ideological objections to working within Intelsat's commercial structure, this was not really the case.[48]

At the same time that it wooed Western European members with offers that closely resembled those countries' demands toward Intelsat, Interkosmos also sought to tempt future Eastern European Intersputnik members with access to manufacturing contracts. Like their Western European counterparts, Soviet-bloc countries sought access to high tech manufacturing contracts. During the summer of 1968, in the months before a late-June meeting in Budapest where the Intersputnik proposal was to be finalized, the Soviet Ministry of Communications signed agreements to conduct joint research and component production with the German Democratic Republic (GDR), Bulgaria, and Czechoslovakia.[49] Interkosmos's Soviet leadership also explicitly promised its Eastern European participants that Intersputnik's prices would be lower than those of Intelsat for the same services.[50] In the summer of 1969, moreover, a Soviet diplomat told a US counterpart in Geneva that in the network's first phase, network members would be granted the use of Soviet satellites without charge.[51] Thus, while US diplomats tended to characterize socialist countries' decisions to join Intersputnik as entirely political, economic incentives were also an important part of the negotiations within the socialist bloc in advance of the Vienna announcement.[52]

The early phases of Intersputnik's creation were thus characterized by significant mimesis of Intelsat's structure and were directly shaped by the contemporaneous renegotiation of Intelsat's governing structure. Soviet communications officials were mindful that Intelsat had taken years to negotiate and build: their network, they cautioned, would likely require a similar amount of time to develop.[53] But this was not simply a case of political or economic imitation, or of competition between two separate, opposing networks. Instead, the boundaries between the two networks were initially not firmly set at all, and indeed the possibility of Soviet-bloc entry into Intelsat remained open well past February 1969. Instead of two separate, opposing networks, negotiations in the late 1960s revealed that integration was fundamental to how all sides envisioned the relationship between Intersputnik and Intelsat.

COMPETITION OR INTEGRATION?

Internal discussions on both the US and Soviet sides reveal that some form of integration of the Soviet domestic satellite network, Molniya, with Intelsat was always the expected outcome on both sides. The question was not whether this integration should be accomplished, but on what terms. Even before the Intersputnik proposal, no one on either side seems to have considered the possibility that the Soviet domestic satellite network would not eventually be linked in *some* way to a global system (i.e., Intelsat). As a March 25, 1968, report by the US State Department Bureau of Intelligence and Research pointed out, the integration of the Molniya system into Intelsat's network, without Soviet membership, was likely the most desired outcome on the Soviet side. Such an arrangement would allow the Soviets both political and economic benefits: it could "stay out of what it may feel is a US-operated club, yet at the same time plug Molniya into a world hookup and accordingly enhance its international standing and earnings."[54] The State Department was also open to this possible arrangement; on November 25, 1968, it informed its embassies that it saw no objections to allowing non-Intelsat members "direct access" to the system, provided that financial terms were set so that nonmembers did not have an advantage over members.[55]

Soviet negotiations with likely Intersputnik member-states also made clear that integration and exchange with Intelsat would be central functions of a new, Soviet-led network. In the summer of 1967, Interkosmos sent out surveys asking, among other questions, whether and when each potential member-country would require access to the Intelsat system.[56] Several of the respondents clearly saw Intersputnik as a route to greater global integration and exchange, particularly of television programming. In the fall of 1967, Polish representatives proposed new text for the agreement, including that "the technical parameters of the International System of Communication Satellites must take into account the possibility of cooperation with other systems [in 1967, this could only mean Intelsat], creating the conditions for the future organization of a single global system, accessible to all countries"—a striking echo of President Johnson's invitation to the socialist world to join Intelsat.[57] When the Intersputnik proposal was finalized in June 1968, it included a chapter on cooperation with "capitalist-country satellite systems," which concluded that it would be useful to include "one or two rebroadcast stations [retransliatsionnye stantsii] that will be able to work simultaneously with multiple satellite communications systems."[58]

At the same time, several future Intersputnik founding signatories actively considered joining Intelsat both before *and* after the formal signing of the Intersputnik proposal in 1971. Hungary and Romania were particularly active in their overtures. In July 1968, US representatives were invited to visit Bucharest to discuss Intelsat membership with Alexandru Spatari, president of the Romanian Commission for Aerospace Activities; a trip by Spatari to Washington to meet with COMSAT officials about constructing an Intelsat Earth station was also broached later that fall.[59] Indeed, Romanian interest in building an Intelsat Earth station continued and reached the level of meetings in Bucharest with COMSAT officials and representatives of another US aerospace firm, GT&E International Systems, by 1973.[60] Romania built its Intelsat station in 1976, despite the fact that the neighboring Yugoslavia had joined Intelsat in 1970 and built an Intelsat Earth station shortly thereafter.[61]

Even the very creation of Intersputnik as a separate network remained uncertain and dependent on a variety of other factors for several years after it was announced. For Soviet diplomats in the late 1960s, creating a separate international network was *less* important than the larger goal of integrating a Soviet network with Intelsat. This became evident in conversations within Interkosmos when the hoped-for interest in joining Intersputnik from countries outside the Soviet bloc failed to materialize. As of December 15, 1968, an internal memorandum reported, thirty-two countries (beyond the founding members firmly within the Soviet bloc) had received the Soviet Intersputnik proposal, and not one had yet replied. This could be explained, the report continued, by the fact that many of these countries were already heavily invested in Intelsat; their future relationship with Intersputnik depended on what happened at the forthcoming Intelsat negotiations, set for February 1969.[62] The memorandum went on to outline several possible next steps, depending on how many, and which, countries ultimately decided to join Intersputnik. If a large number of countries agreed to do so, the network could move forward. The same was possible, the report continued, if, for example, only "a small number of countries wish to join," but if they included "countries with significant scientific and manufacturing resources." It could also proceed if a larger number of developing countries, which lacked resources to invest but would "agree to rent channels on the satellite," decided to join. In other words, so long as there were wealthy capitalist world coinvestors *or* there was a clear market for Intersputnik's services in the developing world, plans for the Intersputnik network could go forward after March 1969, reflecting a planning process that was chiefly about economic viability.[63]

However, even if Intersputnik were not created, the memo confirmed that "steps would be taken to develop possible forms of cooperation between the international system 'INTELSAT' and [the Soviet] regional communication system with geostationary satellites." This cooperation could take place via "mutual use of communication channels, or the acceptance of traffic from countries belonging to Intelsat by [socialist-bloc] earth stations and the further transmission of this traffic via land lines to countries that are not members of Intelsat." "Other forms of cooperation could be possible as well," the memo concluded.[64] In effect, the Soviet Ministry of Foreign Affairs position in advance of the Intelsat negotiations of February 1969 was almost entirely flexible and contingent; the only consistent part of their plan was the goal of somehow integrating the Soviet domestic satellite system with Intelsat.

THE US RESPONDS TO THE INTERSPUTNIK PROPOSAL

Soviet flexibility on the specific form of satellite network integration naturally reflected the country's weak negotiating position in the face of an

Intelsat network that was already well established and gaining new members rapidly. Despite the failure of the Intersputnik proposal to attract a significant number of capitalist- or developing-world members, however, it did have an impact within the US government, and ultimately on the governing structures of Intelsat itself. In an August 1968 memo to Secretary of State Dean Rusk about the Intersputnik announcement in Vienna, Assistant Secretary of State for Economic Affairs Anthony Solomon acknowledged that the Intersputnik proposal was "structurally similar to the existing INTELSAT arrangements, except that it provides for decision making in a Council with voting by one-country/one-vote." In response, he proposed that the State Department consider reviving its 1967 proposal to create an annual assembly within Intelsat that would have "quite limited powers" but vote on the basis of one-country/one-vote. "It is my belief," Solomon continued, "that so long as the assembly is not transformed into a body making basic commercial or systems decisions, we can and should be prepared to make the voting in the assembly simply one-nation/one-vote."[65] The memo that Rusk sent to President Johnson a few days later likewise concluded that, looking forward to the Intelsat negotiations in February 1969, "we should be prepared to make such changes in the structure as are necessary and acceptable to continue the very broad support this organization has built."[66]

As the February 1969 negotiations drew closer, US officials continued to note strong desire among some member-countries to reduce US influence, including by eliminating the weighted voting of the interim agreements in favor of a one country/one vote structure and ending US veto power over decision-making within Intelsat. This move would allow the creation of separate regional networks, which the US worried would do "economic damage" to Intelsat.[67] These member-states' positions, White House staff assigned to telecommunications felt, would be strengthened by the Soviet Union's presence as an observer at the February 1969 negotiations. The Soviet presence in the room, White House advisors noted, will "strengthen the resolve of some others, such as France, Sweden, and India, to press their case for lessening US influence." The report continued, "These countries will argue that if the Soviet Union is ever to participate—which is desirable—then the global system must reflect less U.S. influence."[68] Moreover, White House advisors recognized the alignment between the changes that the Soviet Union would likely require to join Intelsat and those desired by European states and other member-states.[69]

When negotiations failed to produce a satisfactory agreement among Intelsat members in the February 1969 meetings, moreover, White House officials continued to discuss Intersputnik's overtures to European Intelsat members as a threat. In a February 1970 memo to Henry Kissinger, for example, National Security Council advisor Helmut Sonnenfeldt noted that the US's firm position on retaining COMSAT as the manager of Intelsat for five years after the signing of new permanent arrangements would be a source of "discord" with European allies that "can become troublesome." This rift could mean, he added, that "the Soviets, over time, may be able to make headway with their 'more democratic' alternative."[70]

The combined pressure of Western European Intelsat members and the Soviet Intersputnik proposal did ultimately lead the US to accede to European demands. After two years of discussions, the Intelsat permanent arrangements signed in April 1971 included the creation of an assembly, focusing on general policy and long-term objectives, and a meeting of signatories, addressing technical, operational, and financial matters, both constituted on a one-country/one-vote basis.[71] COMSAT and US representatives from the White House managed to ensure that COMSAT retained significant decision-making authority in its role as system manager for another six years after the signing of the permanent arrangements; both new one nation/one vote Intelsat governing bodies were limited to an advisory role.[72] Nonetheless, the permanent arrangements reflected a significant diminishment of the US role within Intelsat. However partial, the devolution of power to member-governments under the new permanent arrangements marked the beginning of the end for the US vision of a single global satellite communications network under US leadership. The new permanent arrangements authorized the creation of regional satellite networks by Intelsat member-countries, so long as they consider the recommendations of the Intelsat Assembly with regard to any "adverse financial effect" of a proposed new satellite system on Intelsat.[73] That assembly was, again, the new one-nation/one-vote body comprising all Intelsat members that Solomon had urged Rusk to consider in response to the Intersputnik proposal back in 1967. In effect, this hard-won European demand, strengthened by the threat, however weak, of a Soviet alternative to Intelsat, opened the door for the creation in the 1970s of a wide variety of regional networks, including Europe's Eutelsat and the Arab world's Arabsat, alongside Intelsat and Intersputnik.

"INTELSAT SHOULD HAVE TAUGHT US A LESSON": INTELSAT'S
PERMANENT ARRANGEMENT NEGOTIATIONS AND THE POST-APOLLO
SPACE SHUTTLE PROJECT

Alongside the threat of a rival Soviet network, there was another source of
the weakening US position within Intelsat in the summer of 1971: the fact
that US officials were negotiating with Western European officials about
both the Intelsat negotiations *and* a possible new program of international
cooperation in other forms of space activity at the same time. As the Intelsat permanent arrangement negotiations unfolded in 1969–1971, European
negotiators were quick to link their demands for Intelsat's governance structure under the new permanent arrangements to their financial participation
in a post-Apollo space shuttle program, which negotiators with the National
Aeronautics and Space Administration (NASA) were tentatively exploring in
1970–1971 (well before congressional approval of such a program was secure).
As the National Security Council official Charles Joyce put it, the Europeans
wanted a "more influential role in space ventures" with the US and saw
Intelsat as one arena for pursuing this.[74] In January 1970, the US embassy in
London reported that a West German official named Mr. Brunner, who was
personally involved in both the Intelsat negotiations and in discussions
about European international cooperation in space with the US more generally, suggested that "failure on the part of the U.S. to yield on the matter of
Intelsat control may place obstacles in the way of European cooperation in
our post-Apollo space program."[75]

The White House official Robert M. Behr rejected this European position
as a "crude form of arm-twisting," but the threat gained specificity as both
Intelsat and space cooperation negotiations advanced in 1970–1971.[76] By
early 1971, both the Federal Republic of Germany and France were insisting
on a link between the question of European financial and technical contributions in post-Apollo space cooperation and the issue of US launch guarantees
for European satellites. The US, bound by both Intelsat's own rules and the
desire to ensure Intelsat's monopoly as a "single global system," sought to
reserve the right to refuse to launch communications satellites that would
pose a competitive threat to Intelsat. The French and West Germans insisted,
in turn, that they could not invest in post-Apollo space cooperation if they
could not be certain that the US would launch their satellites; without such
confidence, they would have to invest their resources in developing their

own launch capacities.[77] US efforts to craft a response that would reassure Europeans that a refusal to launch a European satellite was very unlikely were complicated by the fact that decisions about whether new satellites in fact posed a competitive threat were to be made by Intelsat's governing body, the structure of which was also under intense negotiations in the spring of 1971. If the US were going to respect Intelsat's autonomy, it could not guarantee the French that future Intelsat decisions would always support French goals.

In effect, the US found that its two major cooperative engagements in space after Apollo—Intelsat and the planned shuttle program—were in direct conflict with one another. Kissinger saw the international prestige that the US gained after Apollo 11 as an opportunity to strengthen alliances with European partners and US centrality in an envisioned new space transportation infrastructure. Equally important was the fact that a substantial European financial contribution to a future space shuttle program could weaken criticism that investment in space after Apollo 11 was an unnecessary cost. But it was precisely this financial contribution that allowed France and other European powers to argue that they could not move forward with cooperation in space if their demands for launch guarantees for European satellites and changes to Intelsat's governing structure to weaken US dominance there were not met. France in particular was committed to the creation of a European satellite network that was not under US control; moreover, it would not settle for language about regional networks being permissible within Intelsat's rules because it did not wish to exclude the pursuit of a French-led satellite network uniting the former French empire around the globe.[78]

White House and NASA officials differed over the question of what kinds of concessions were acceptable in order to make European participation in the shuttle program happen. In this context, the ongoing US concessions to the French and other European parties in the Intelsat negotiations served as a cautionary tale. In an April 23, 1971, meeting, one White House staff member, Tom Whitehead, urged officials to make sure that the US's own interests would genuinely be served by the terms of post-Apollo cooperation, rather than once again offering the Europeans disproportionate benefits, noting that "our experience with INTELSAT should have taught us a lesson."[79]

Perhaps because of this sense that maintaining a strong US position within Intelsat was already a lost cause after the April 1971 signing of the new Intelsat permanent arrangements, as negotiations continued, US representatives ultimately decided to significantly weaken both COMSAT's position within Intelsat and Intelsat's monopoly on global satellite communications in order to pursue other forms of international cooperation in space. Doing this opened the door to competitors to Intelsat in the form of regional and even global networks created by some Intelsat members over the wishes of others. From NASA's perspective, articulated in a report on "technology transfer in the post-Apollo program," offering Europe unlimited launch assurances for communications satellites that would make up these new networks was preferable to seeing European countries develop their own, rival launch capacity for satellites and other payloads.[80]

Correspondingly, in the ongoing internal discussions in the White House, the issue of legal compliance with Intelsat's rules regarding members creating or joining other satellite networks was increasingly seen as an obstacle to be got around. In a July 27, 1971, letter from Kissinger to Secretary of State Bill Rogers, Kissinger suggested a solution to the impasse with Europeans in the post-Apollo talks: First, as before, the US should agree that it would launch any foreign payloads that were peaceful in nature and approved by Intelsat. However, if necessary, Kissinger suggested, the US might also promise to sell the "necessary launch vehicle" to countries wishing to launch unapproved satellites, "leaving to the launching nation the interpretation of its obligations" under Intelsat's bylaws.[81] Given the singular power that the US wielded within Intelsat, if it did not support constraints on Intelsat members' creation of rival satellite networks, who would? For Kissinger and for NASA, the expansion of post-Apollo space cooperation with Europe was more important than supporting the pursuit of a single system for satellite communications.

CONCLUSION

Perhaps the most striking evidence that the changes to Intelsat's organizational structure that resulted from all these pressures and competing priorities were meaningful was their reception by Soviet specialists. In a report submitted to the Communist Party Central Committee's General Section in early February 1971, I. V. Vasilieva, a researcher in the Soviet Academy of Sciences Institute for Applied Social Research, carefully built a passionate, if cautious, argument for Soviet membership in Intelsat. Vasilieva began by outlining Intelsat's and the US's enormous technical and political head start in building an operational global satellite network, as well as the significant ideological danger of allowing US programming to flow unchecked into the postcolonial world. Most important, Vasilieva stressed, was how very different Intelsat's governance structure now was, and how much closer to Soviet negotiating positions in the late 1960s. The draft Intelsat Permanent Arrangements to be signed that April had changed from the interim agreements "in the direction of greater democracy, increased rights for member countries, greater limits on countries with large financial contributions, access to the space segment for countries who are not INTELSAT members, and so on." Soviet membership in Intelsat would only further weaken US power within the organization, Vasilieva argued, and Soviet technology "would have a colossal new export market."[82] Fantasizing about Soviet television programming flowing freely around the world over rented Intelsat channels, she stressed that the real problem would be how to produce enough content in the relevant languages and based on a scientific knowledge of local desires and tastes.[83]

Vasilieva's proposal that the Soviet Union join Intelsat under its new permanent arrangements of April 1971 was not adopted.[84] But in a larger sense, it did not have to be. While the free broadcast of Soviet television around the globe on rented Intelsat channels did not come to pass, the circulation of socialist world television programming via satellite, albeit on a much smaller scale, was realized via Intersputnik just a few years later, adding its small part to the rather robust world of Soviet global cultural exchanges with other socialist and nonaligned countries, particularly in radio, film, and print media. Furthermore, Soviet full membership in Intelsat was not necessary. Thanks to the alignment between Soviet financial and political interests and those of many European and some postcolonial members of Intelsat, the Intelsat negotiations of 1967–1971 had already accomplished the Soviet bloc's key goals.

The ground was laid for the fragmentation of satellite communications infrastructure into a pluralistic mix of overlapping regional and global networks, reducing US power and preventing a monopoly by a US-dominated Intelsat. Moreover, the Soviet Union and its allies gained access to Intelsat's network and television flows via the Intelsat Earth stations that were constructed in multiple socialist world countries by the mid-1970s, as we outline in chapter 5. As Intelsat Earth station owners, the Soviet Union, Romania, Czechoslovakia, and others became participants in Intelsat's annual conferences for Earth station operators, exchanging technical information and building international professional contacts.

The signing of two, seemingly separate and rival, satellite communications network agreements in 1971 thus did not prevent, and indeed even facilitated, interaction, exchange, and gradual integration across the Iron Curtain in the context of a competitive system of multiple regional satellite networks. The overlap between Soviet and Western European commercial as well as political interests, rather than unilateral US scientific and commercial superiority, reshaped the institutional structures that would underpin commercial communications satellite organizations for the next two decades and beyond. Despite the Soviet Union's genuine economic and geopolitical weakness relative to the US in the postwar decades, the Soviet Union and its Warsaw Pact allies actively sought integration while working with Western European governments to reshape communication infrastructures and institutions, reducing US dominance and driving economic and media globalization forward in pursuit of shared goals.

Just as the two satellite spectaculars discussed in chapter 2, "Our World" and "One Hour in the Life of the Motherland," reconceptualized and recalibrated the planetary in light of competing temporalities of the broadcasts, so did the institutionalization of communications satellites and the networks that they eventually formed contribute to planetary thinking. What was at stake was not only the question of belonging to Intelsat or Intersputnik, but also fantasies about what such belonging would entail, be it equal power and influence, access to global audiences, or entry into new global markets. The creation of governance institutions for global satellite communications thus suggests how incomplete an exclusively binary, competitive framework is for understanding the history of technology during the Cold War, as well as how infrastructural fragmentation—the creation of multiple, regional satellite networks instead of a single global network—can contribute to, rather than impede, globalization and new, planetary ways of thinking.

4 "SPACE BEGINS ON EARTH": SELLING, BUILDING, AND REPRESENTING SATELLITE EARTH STATIONS

On October 6, 1969, Abbott Washburn, the US representative to the Intelsat negotiations, presented an update on the course of those negotiations to an audience at a seminar organized by the Electronic Industries Association on the topic of "Satellites and Sales—Impact on International Electronic Business." Despite the ongoing challenges of those negotiations, Washburn opened his address on a triumphant note, with a snapshot of Intelsat's global expansion. "Last Thursday," he reported, "Argentina's earth station was inaugurated" and "a week ago last Tuesday the Hong Kong earth station initiated service." He continued, "Iran came in over this weekend, on Saturday, October 4." Depicting a rapidly expanding network, in which the opening of a new Intelsat Earth station took place every few days, Washburn's opening remarks reflected not only optimism, but also the importance of Earth-based infrastructures for satellite communications—satellite "Earth stations"—to the construction of Intelsat's global network. Washburn insisted to his audience that what distinguished satellite communications from undersea cables was the former's ability to "provide multiple access to every country looking at the same bird, whereas cables are essentially a point-to-point service." Yet what Washburn described as Intelsat's "global coverage," via five geosynchronous satellites, was meaningless without the terrestrial network of Earth stations that were needed to receive and distribute their signals to existing telephone and television networks on Earth. Washburn concluded his remarks by noting this dependency. "Earth stations," he explained, "is one area over which INTELSAT, as such, has no control," since they are "procured, built and maintained by the appropriate authorities within the country where the facility is located."[1]

As Washburn's remarks indicate, the fraught negotiations over the future structure of global satellite communications were not only understood in abstract terms of future profit, industrial growth, Cold War rivalry, and membership in a growing high-tech economy. These global rivalries were also embodied materially in the race to construct earthly infrastructures for heavenly satellites. While participation in the space segment of satellite communication systems was ultimately controlled by the two superpowers that monopolized launch capacity in the late 1960s, the ground segment—a network of Earth stations that could receive satellite signals and redistribute them over local and regional cable and microwave networks—was more open.[2] US, socialist state, Japanese, and European corporations and state agencies could all bid to build satellite communications Earth stations around the world, and any country could, at least in theory, become the proud owner of one of these new symbols of Space Age modernity. Washburn's confession that Intelsat could not fully control its ground segment reflects the disjuncture between the coverage zone of the satellite, the part of Earth's surface where a satellite's signal can potentially be received, and the terrestrial infrastructure, equally shaped by geopolitical forces, which must be built in order to receive it. Creation of this terrestrial infrastructure was thus an essential part of making satellite communications truly global.

Just as photographs like "Earthrise" and "Blue Marble" purported to show the entire planet while by necessity actually displaying only a fraction of the globe, the promotion and expansion of early satellite networks were structured by a tension between the claim of global access and presence and the reality of situated, inherently limited networks. This tension carried through the stages of global ground station construction that we trace in this chapter, from the promotion and selling of satellite networks as global through the highly localized process of negotiating and building specific Earth stations. Throughout, the effort to sell and build satellite communications ground structures around the world, as well as the technical features of early satellite Earth stations themselves, threatened Cold War boundaries and complicated US efforts to maintain Intelsat's dominance within this emerging communications market.

Since they were remote, like satellites themselves, Earth stations had to be explained, celebrated, and generally made visible as part of the promotion

and sale of satellite communications to the state telecom officials and, ultimately, global citizens who were asked to support investment in the construction of an expensive and not immediately fiscally self-sustaining medium.
The result was the wide circulation of photographs of the new Earth stations
in postcards, popular-scientific films, and even postal stamps issued by the
countries that built them. These images deployed a particular image of space
as both apolitical and, at the same time, discretely national in order to efface
questions of power and fears about the fluidity of Cold War alliances and the
penetrability of political borders. The effort to sell and build communications satellite ground infrastructure, we find, thus had a double impact, both
promoting media globalization and concealing it from public view.

SELLING EARTH STATIONS

The late 1950s saw the advent of experimental satellite communication systems, most notably Project SCORE (whose name stands for "Signal Communications by Orbiting Relay Equipment"), a military initiative using a ballistic
missile to launch a satellite that communicated with four ground stations in
the southern parts of the US.[3] With the launch of Telstar in 1962, facilitating
the first transatlantic telecast, communications satellites shifted from being
hidden and experimental to being actively publicized and widely known to
the general public.[4] Despite using only three ground stations, in the US, UK,
and France, respectively, the Telstar experiments have been seen as a first step
toward a global satellite communication system and what has been called
"informational globalism."[5] Yet, as Washburn's October 1969 remarks suggest, there could be no truly global satellite communications system without
the construction of Earth stations around the globe.

For the telecommunications industry, this meant the emergence of a lucrative new market. By the mid-1960s, with support from the US State Department and as part of efforts to ensure that AT&T did not monopolize this
new telecommunications medium, RCA and Hughes were granted satellite
contracts with the National Aeronautics and Space Administration (NASA),
and RCA began to actively promote its services as a designer and builder of
satellite Earth stations.[6] A glossy 1964 RCA pamphlet entitled "Ground Stations for Space Telecommunications" presented the need for action on the
part of national telecommunications officials as urgent: officials, the pamphlet urged, "must study their own requirements, and estimate probable

expansion of demand in the next 5 to 10 years," and determine "how their ever-increasing traffic may most effectively make use of satellite communications."[7] Although commercial satellite service was not expected to begin for another year at least, RCA urged that governments would not wish to be left behind, unable to deal with what the company presented as their ever-increasing communications traffic.

RCA's urgency, of course, had more to do with its own interest in expanding into this new market than with the immediate need for satellite communications in most of the world.[8] Yet its marketing materials offer a view of the effort to link the construction of satellite Earth stations firmly to membership in the US's Cold War camp. RCA's marketing materials informed buyers that their considerable technical experience and assistance was available to "any administration in the free world"; moreover, they hinted that that expertise extended far beyond civilian satellite communications to include RCA's substantial experience building US military communications infrastructure.[9] The brochure's pitch to readers began with a list of RCA's previous achievements in satellite communications, including collaborations with both NASA and the US army. RCA was offering, the brochure stressed, all of "its knowledge, its experience, [and] its research and manufacturing facilities" to interested countries. In case readers missed the point, the brochure then went on to list the RCA facilities to which clients would gain access. These included "the RCA Defense Electronic Products organization, including the Astro-Electronics Division, Aerospace Systems Division, Missile and Surface Radar Division, and Communications Systems Division."[10] Moreover, although the Soviet Union was, in 1964, still only planning a broader network of domestic ground stations, with plans for construction and/or sale of its Earth stations beyond Soviet borders still distant, the RCA pamphlet reflected an eagerness to denigrate the Soviets' role and exclude them from this new sector. The RCA brochure boasted that its scientists had been at work imagining and planning for satellite communications since well before Sputnik's successful 1957 orbit, downplaying the significance of that Soviet Space Race "first."

This blurring of lines between civilian and military technical assistance was further reinforced by the brochure's illustrations, which included numerous photographs of military satellite communications ground stations, including an unidentified photo of a "typical ground station of the U.S. Ballistic Missile Early Warning System [BMEWS]," a second photo of a BMEWS ground station in Moorestown, NJ, and a photo of a "typical high-power microwave

FIGURE 4.1
Ladislav Sutnar, rendering of a planned Earth station in Nova Scotia for RCA. Reproduced with permission, Radoslav Sutnar.

antenna and pedestal, designed by RCA for military purposes." These military installations, of course, were the only existing RCA-built satellite ground facilities in 1964, and thus the only real source for the brochure's photo illustrations. Yet a striking artist's rendering of Canada's planned Earth station in Nova Scotia likewise offered the brochure's readers a chance to penetrate what were otherwise secret and invisible government technical installations (see figure 4.1).[11] To show what precisely was inside the Earth station's protective "radome" (a weatherproof enclosure, maximally penetrable by radio waves, that protects radio equipment from weather and other damage), the artist drew the radome with a section broken away, revealing the large satellite dish inside.

Despite these beguiling promises of access to what had been chiefly US military technology until recently, the notion of US dominance of this new economic and technical sector was also potentially alienating to other countries. RCA's brochure thus had to frame its relationship to the US government, as well as its emphasis on US technological dominance in this sector, in ways that would actually appeal to national telecom officials rather than alienate them. The brochure's text and illustrations accomplished this in several ways. First, as already discussed, the brochure made an economic modernization argument that, without access to satellite communications, countries would find their current communications infrastructures overwhelmed by increasing traffic demands.[12] This argument cleverly drew on economic and ideological arguments about the nature of global economic growth, grounded in technocratic expertise. To counter the brochure's argument, national governments would have to argue that their countries, however small or burdened by postcolonial political and economic challenges, would be excluded from dramatic economic growth in coming decades.

Maps depicting the coverage zones of Intelsat satellites contributed to this sense of urgency by conveying to potential Earth station–building countries that a global satellite communications system was already an accomplished fact. One of these coverage maps were published in a promotional folder from COMSAT depicting the construction of the Earth station in Andover, Maine.[13] Here, the coverage zone of Intelsat II was depicted with a small number of European and US Earth stations already in operation marked on it, together with a much larger number of planned stations in South America, Africa, and the Middle East, and even in India, well outside the reach of Intelsat II (figure 4.2). The inclusion of the planned stations conveyed not the present, but rather an imagined future of increased range and connectivity; as the map's description explained, the "system [was] being expanded to a global scale."

This map depicts satellite signal coverage as undifferentiated, suggesting that any point within the coverage zone would have equal access to the satellite's channels. Like the language used to promote Intelsat's space segment as offering "total global coverage," "planetary coverage," and working on a "truly global scale," coverage maps obscure the fact that they depict only the potential maximum reach of the system.[14] In the pre-direct-broadcast era, as already mentioned, reaching all the areas of Earth's surface depicted required the construction of specialized Earth stations in those coverage zones, as well as linking them to new or existing radio relay networks that would distribute the signal on the ground.[15] While these maps often did include representations of existing and planned Earth stations within the coverage zones, the ground infrastructure beyond the Earth station—the real horizon for signal

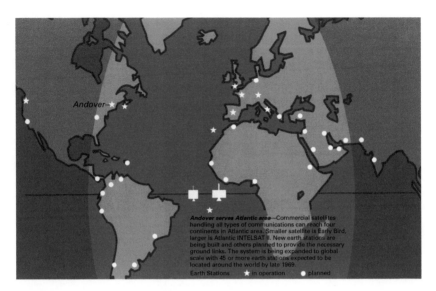

FIGURE 4.2

Coverage map, Intelsat II, 1967. COMSAT, "Andover Earth Station," ca. 1967 (Washington, DC: COMSAT Information Office), 2.

distribution beyond the Earth station—was never included. These maps *also* concealed the institutional channels through which flows of content via the satellite were negotiated and paid for, which would, at least in theory, constrain access to satellite signals on the ground. The reality of satellite failure, which plagued Intelsat in its early years (as when an antenna on the Intelsat III satellite over the Atlantic failed in June 1969, rendering the satellite unusable), was also not reflected in these coverage maps, of course.[16]

These Intelsat coverage maps, designed to attract new member-countries and sell Earth stations, had another notable feature: they rendered the Soviet Union invisible as a rival and alternative source of communications satellite service. While, like the exclusion of satellite technical failures, this is somewhat unsurprising, given the promotional intentions of coverage and Earth station maps, it is nonetheless significant. First, as noted in chapter 3, many Intelsat member-countries sought to gain Soviet entry into Intelsat for a variety of political and commercial motives. Second, the ways in which these maps, as a group, render the other space superpower invisible are rather dramatic. Many Intelsat coverage maps used a projection that placed North America and Africa at the center of the world map, dividing the Eurasian

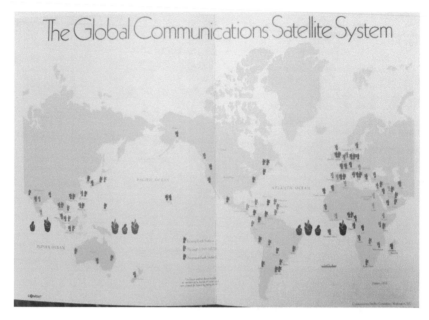

FIGURE 4.3
COMSAT, "The Global Communications Satellite System," February 1971.

landmass and moving it to the periphery.[17] On these maps, the Soviet Union's
territory is typically depicted as a vast, empty space without either Earth stations or satellite coverage (despite the existence and active operation since
1967 of the Molniya and Orbita domestic satellite broadcasting systems), as
in the February 1971 map of the global communications satellite systems
depicted in figure 4.3.[18]

Similarly, a 1969 press release from the Indonesian Satellite Corporation,
touting the "vast span" of Indonesia's planned Intelsat Earth station once
it became operational, placed an enormous drawing of that station directly
over Soviet territory, covering most of central Asia and western Siberia—
an area that in fact had been full of Soviet Orbita Earth stations by 1969
(figure 4.4).[19] Like most US-made Intelsat maps, this Indonesian map renders
Soviet territory as a blank space, and thus a convenient spot for a large illustration of the Earth satellite station in Djakarta.

Soviet illustrations and maps of the initial Molniya satellites and the Orbita
ground network shared several key features with Intersputnik's footprint and
other promotional maps, despite the fact that the Soviet illustrations were

P.T. Indonesian Satellite Corporation

PRESS RELEASE

FIGURE 4.4
The Indonesian Satellite Corporation's rendition of its prospective Earth station.

aimed at informing and impressing domestic audiences, rather than at selling Earth-station technology or network membership to international government telecommunications officials. An illustration in a Soviet popular
science magazine, *Tekhnika Molodezhi* (Technology for the youth), shows
Molniya I in its elliptical orbit above the Soviet Union at a distance of 39,380
kilometers (figure 4.5).[20] Back on Earth, the signal is beamed from a tall tower
in Moscow to be relayed by the Molniya satellite back to Vladivostok. This
artistic, rather than schematic or scientific, image thus emphasizes a point-
to-point connection, not a coverage zone or footprint across the entire Soviet
Union. The vast area between Moscow and Vladivostok, to be connected via
satellite, is divided into two significantly different parts. The area west of the
Urals is scattered with a network of broadcast towers in a star-shaped formation, demonstrating how that part of the Soviet Union was already linked
together by a terrestrial communications network, whereas east of the Urals
is a vast empty space, with no indication of either existing infrastructure or
people being there. Made invisible here is not only the Molniya satellite's
footprint outside Soviet borders, but also the handful of tracking stations

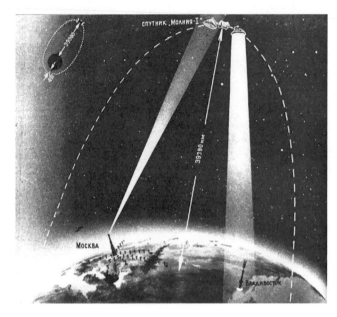

FIGURE 4.5
Molniya I illustration in *Tekhnika Molodezhi*, July 1965.

in the rest of the Soviet Union that served as ground stations for the new Molniya satellite's television broadcasts. In this 1965 illustration, when those tracking stations were still classified as scientific and military facilities, the Vladivostok television tower appears to be receiving Molniya's signal.

The construction of a dedicated set of Orbita communications satellite ground stations in the second half of the 1960s meant that the ground infrastructure of the Molniya satellites was no longer classified. A 1969 illustration (figure 4.6) by A. Minenkov in *Aviation and Cosmonautics*, the professional journal of the Soviet air force, included a map of Orbita ground stations, while foregrounding the satellite itself and its connection with an Orbita Earth station.[21] While the illustration is artistic rather than exclusively technical, just as the one in *Tekhnika Molodezhi*, the technical details are vivid and the circular bases of Orbita Earth stations are easily recognized. Compared to the earlier image, the landmass of the Soviet Union is now covered with twenty-three Earth stations, from Murmansk in the west to Vladivostok in the East. The footprint of the Molniya satellite in the upper left, however, appears to be limited to Soviet territory, delineated with a sharp red boundary line.

FIGURE 4.6
Illustration of a Molniya satellite and Orbita Earth station by A. Minenkov.

Just as the Indonesian map concealed the Soviet Union's territory with
a large illustration of an Earth station, here neighboring countries are covered by clouds. Moreover, like the RCA brochure's image of an Earth-station
antenna with the radome peeled away, this 1969 illustration connects the
people and landscape of the Soviet Union with outer space. The image's heavenly upper half features a mysterious black sky dotted with bright stars and
even a comet, whereas the lower half contains a silhouetted image of winter
landscape, snow-covered trees, and someone traveling on a sleigh pulled by
reindeer. Satellite communications, this image proposes, had connected the

indigenous peoples of Siberia to modern Space Age technology. However, the cultural and place-based specificity of this particular indigenous sled driver is entirely erased in favor of a stereotype; the viewer is left to wonder which of the tens of satellite Earth stations depicted scattered around the Soviet Far North, Siberia, and Far East is the one in the illustration. This image thus instrumentalizes indigeneity to highlight the supposed contrast between the traditional transportation of reindeer herders and the hypermodern communications satellite Earth station. At the same time, the beautiful twilight landscape and exquisite color in the artist's illustration aestheticize not only indigenous people and landscapes in the Soviet far north, but also planet Earth as a whole, which appears here banded with the colors of sunrise against the darkness of space.

Together, these two illustrations suggest that, despite the domestic focus of the Orbita system, Soviet efforts to explain and promote this new communications system shared key themes with those produced by Intelsat-affiliated corporations and governments. Those included the exclusion and erasure of territories supposedly not connected by satellite, an emphasis on the ways that satellite ground infrastructure could connect modernizing countries and regions with Space Age technology, and a tendency to depict Earth in its planetary context, in dialogue with space beyond its surface.

BUILDING EARTH STATIONS

For RCA, Intelsat, and other developed world firms and organizations, selling Earth stations to national telecom agencies was not a straightforward process. Despite promotional materials and illustrations that erased the Soviet Union and its allies from the map, it was not so easy to entirely ignore the existence of the socialist world in determining the location of future Earth stations. Neither RCA nor any of the many other aerospace firms interested in building Earth stations could simply sell Earth stations to any country in the so-called free world without paying any attention to local and regional contexts and shifting Cold War geopolitics. Indeed, the entanglement of the economic and geopolitical became more explicit as the intertwined processes of recruiting new Intelsat members and constructing Earth stations in their countries accelerated in the second half of the 1960s.

In May 1966, collaboration between the US State Department and RCA culminated in a seminar in Washington, DC, on Earth-station construction that welcomed fifty-nine representatives of thirty-nine countries and included a trip to tour the recently completed Nova Scotia Earth station.[22] This seminar was part of a broader US effort to recruit a priority group of thirteen so-called less-developed countries to join Intelsat, in conjunction with constructing an Intelsat-aligned Earth station. Overtures to these countries, chiefly in Latin America, Africa, and South and Southeast Asia, were designed to educate them as to "the value of membership in the global system," as well as "the desirability . . . of establishing earth stations."[23] A December 31, 1966, memorandum to Walter Rostow in the White House reporting on progress toward this goal observed that major obstacles to Earth station expansion had been "organizational and technical problems [that] have held up some decision making by the developing countries."[24] Financing had not been a major obstacle to the construction of communications Earth stations in these countries; many were able to self-finance or receive private financing, whether from banks or directly from the firm selected to lead the project.[25]

One source of these delays was the fact that the planning of Earth-station locations was an inherently supranational problem. There were technical questions about which geostationary satellites were visible from a particular point on Earth's surface; Earth stations had to be built inside the coverage zone of the targeted satellite. Also, it was crucial to make sure that the projected traffic from the set of Earth stations sending and receiving data from a particular satellite would not exceed that satellite's capacity.[26] From the perspective of prospective Intelsat members, however, the cost of constructing an Earth station and its likely ability to generate revenue that could cover its construction and ongoing costs were the most important considerations. Earth stations were expensive to build and required elaborate economic projections to assess when (or even whether) they would begin to pay for themselves. How to make these calculations was the subject of research by COMSAT and interested US firms.[27]

One typical paper of this sort demonstrated how best to predict Earth-station economic viability based on traffic projections, using the example of a fictional Latin American country rather uncreatively named "Latina."[28] Beyond the calculations of technocrats, countries themselves had to determine which territories an Earth station might serve and how it would connect to local cable and microwave networks on the ground. Essential to all these calculations was the location of other Earth stations in a given region, since one Earth station could easily handle the needs of several countries, particularly if they were geographically small or had limited traffic needs thanks to either economic underdevelopment or well-developed radio relay and cable networks. The ability of a new Earth station to connect to existing radio relay networks on the ground was also an important factor. A satellite's footprint was meaningless without ground infrastructure to distribute its signal, despite rhetoric about the ability of Intelsat's space segment to reach "distant corners of the world."[29]

As the process of actually building Intelsat Earth stations unfolded, the fundamentally regional and transnational nature of satellite infrastructure construction became even more apparent. The reality of these negotiations was much different from the fantasy depicted in the RCA brochure of US firms, with federal government support, selling Earth stations on a strictly bilateral basis to national telecom agencies.[30] The decision to build an Earth station depended very much on the choices made by neighboring countries and, in divided Europe in particular, tended to raise the question of whether countries could share infrastructures across Cold War boundaries. The problems of negotiating the location of particular Earth stations in specific regional contexts necessarily engaged multiple, conflicting interests and opened the door to competition from alternative regional configurations and rival powers. From 1966, knowledge of plans for a Soviet-led alternative to Intelsat were widespread, opening up the possibility of obtaining an Earth station, likely at a discount, from outside the free world. Moreover, compared to satellites themselves, Earth stations were relatively straightforward to design and build. US firms able to construct Intelsat Earth stations were quickly joined by competitors from Europe and Japan. Having countries finance and build their own Earth stations without US financial assistance was, indeed, an explicit policy, given the strong desire to demonstrate that at least some US space programs offered economic returns that exceeded investment.[31] This came with a trade-off, however—namely, that US firms would be quickly displaced from this lucrative new market, despite their promotional efforts with support from the State Department during the 1960s. As of May 1976, US firms had built only about 20 percent of Intelsat Earth stations (37 out of 183), and the US Department of Commerce expected US competitiveness to decline in the future.[32]

The negotiations leading up to the construction of an Intelsat Earth station in Yugoslavia offer one example of how the construction of satellite ground infrastructure could bring economic and Cold War political logics into conflict. Like their French counterparts, Yugoslavia's government sought to use its position as potential mediators between Cold War rivals and employed the threat of joining a Soviet-led communications satellite system to extract concessions from the US and Intelsat. In early August 1967, a year before the Intersputnik network was publicly announced, the State Department received an airgram from the US embassy in Belgrade, stating that Yugoslavia still not had made up its mind which satellite system to join. The airgram reported on a meeting with two counselors for the Yugoslavian Post, Telephone, and Telegraph administration (PT&T), Konstantin Comic and Dusan Milankovic, who wanted to push back on a local newspaper story that stated that Yugoslavia had made up its mind which network to join.[33] On the contrary, the officials reported, they were still considering the economic viability of joining Intelsat. A key factor in their decision, they insisted, was whether a Yugoslav Earth station would also serve Greece. The Yugoslav officials already felt confident that a Yugoslav Earth station would immediately be able to carry all of Romania's traffic, and likely Hungary's, soon after.[34] The financial viability of a Yugoslav Earth station was thus understood to be based on the ability to serve a number of regional neighbors, without regard to their Cold War alliance status.

The following spring, the plans for a Yugoslavian Earth station and subsequent Intelsat membership began to emerge more clearly. In a meeting on March 26, 1968, General Director Prvoslav Vasiljevic of the Yugoslavian PT&T made it clear that, first, Yugoslavia had not received an invitation to join a Soviet-sponsored system, and, even if it had, that the Soviet Molniya space segment "cannot assume [the] role of truly international system, and thus Yugoslavia would not be interested in joining [Molniya/Intersputnik]."[35] As the officials stressed, constructing an Intelsat Earth station in Yugoslavia would be "tantamount to deciding to join INTELSAT, which is the only international system in being [sic] and is already functioning well."[36] The remaining obstacle, discussed in the balance of the telegram reporting about the meeting, was how to assemble the funds need to construct the Earth station; the message concluded with the observation that "it seems clear that Yugoslavs wish [to] explore whether special financial assistance can be arranged in accord with President's message of August 14, 1967."[37] In mid-January 1970, Yugoslavia officially confirmed its desire to join Intelsat and on April 30, 1970, it formally became a member. Yugoslavia's decision may seem unsurprising, given its independent foreign policy and economic relations. But other socialist countries were equally eager to construct an Intelsat Earth station, even if they did not formally join the network. After a decade of expressing interest, Romania also built an Intelsat Earth station, in 1976.[38]

The process of construction of Intersputnik Earth stations was somewhat less fraught by regional considerations, for several reasons. First, construction of Intersputnik ground stations took place several years after the construction of Intelsat Earth stations was well underway, and it initially was limited to Soviet-bloc allies that had not yet built Intelsat stations. Intersputnik members, before the end of the 1970s, were predominantly countries that belonged to the International Organization for Radio and Television (OIRT), and much of their initial traffic involved the exchange of television programming within the Soviet bloc, a service that was less expensive to conduct over the existing radio relay networks. The first Intersputnik Earth station beyond Soviet territory was built in Cuba, a Soviet ally for whom facilitating television broadcasting exchange and telephony with Eastern European socialist countries genuinely required communications satellites. Moreover, that first Intersputnik ground station, built in Jaruco, Cuba, just outside Havana, in 1973, was constructed not by international telecommunications firms, as had been the case for Intelsat stations, but from components chiefly made the Soviet Union and installed by teams of mostly Soviet engineers. These stations were highly standardized from the beginning, and their distinctive circular shape remains immediately recognizable today (figure 4.7).[39]

This way of organizing ground station construction was not necessarily the product of a distinctively autarkic Soviet approach to construction contracts in the developing world. As Lukasz Stanek and others have documented, socialist world architects who designed and helped build housing and many other landmark public buildings across the Global South in the 1970s and 1980s generally collaborated extensively with local officials and architects, resulting in extensive customization and adaptation of materials to local climatic conditions and cultural preferences.[40] By contrast, the construction of Intersputnik stations around the world by the late 1970s and 1980s seems to have been far more closely directed by Soviet specialists, though with some participation by scientists from other Intersputnik member-countries.[41] In a commemorative volume published by the Special Construction Bureau of the Moscow Energetics Institute (OKB MEI), the scientific institute responsible for design, construction, and installation of satellite antennae and Earth stations for Intersputnik, and featuring biographies of engineers who took

FIGURE 4.7
The Intersputnik Earth station in Psary, Poland. Reproduced with permission, Intersputnik IOSC.

part in Earth-station installations around the world, individual Earth stations appear only as place names in a long list of installation destinations, designed to convey the prestige, and wide travels, of the Soviet engineers profiled in the commemorative volume.[42] More research remains to be done on the installation process for both Intelsat and Intersputnik Earth stations in order to assess how much collaboration took place between local engineers and construction workers and the teams of installers from aerospace firms or the Intersputnik organization.

Nonetheless, based on the sources that we have been able to access, Intersputnik Earth-station components seem to have been predominantly manufactured in the Soviet Union, with some limited components produced by other member-states. This overall picture reflects the energy with which the Soviet Union was pursuing global profits from its space technology sector, including via deals with global aerospace corporations, fitting within recent work that has focused on Soviet participation in the processes of capitalist globalization. Records in the Soviet Ministry of Communications archives suggest that Soviet officials were taking fifteen to twenty meetings a month with multinational aerospace corporations and other capitalist actors by the mid-1980s.[43] This level of engagement with international corporations even before Mikhail Gorbachev's selection as general secretary of the Soviet Communist Party laid the groundwork for the rapid privatization and entrance of foreign capital into the post-Soviet Russian space industries.

This ongoing integration of global satellite communications networks was happening on the ground in Intersputnik and Intelsat member-countries as well. By early 1979, the Cuban Intersputnik station was accompanied by a Cuban Intelsat Earth station being constructed nearby to provide additional broadcasting capacity for the sixth summit of nonaligned countries, which took place in Cuba in September 1979.[44] Similarly, when Intersputnik began to expand its Earth-station network in the late 1970s and 1980s, it frequently built Earth stations in countries that already had Intelsat stations, such as Iraq, Nicaragua, and Algeria.[45] In all these cases, practical, regional, and commercial considerations, as well as the desire of individual governments to participate in this new, Space Age communications network in the specific ways that best suited their needs, were the most significant factors shaping decisions about where Earth stations were constructed. The construction of both Intelsat and Intersputnik Earth stations challenged the idea of the planetary and indiscriminate coverage of satellites and instead demonstrate the highly local and regional decision-making that led to the construction of each new satellite Earth station.

"KEEP THE FACES OF THESE NATIONS TURNED TOWARD THE WEST": THE PROBLEM OF STEERABILITY

This picture of Earth-station development in the late 1970s and early 1980s, where a handful of countries built both Intelsat and Intersputnik Earth stations alongside one another, sharing auspicious geographical locations and critical infrastructure like power and water lines, would have horrified US officials only fifteen years earlier. In the early 1960s, US diplomats and telecommunications experts had expressed fundamental concerns about the unstable relationship between satellite communications infrastructures and

geopolitical alliances. While US officials were initially confident that technical features specific to Intelsat or Intersputnik satellites would limit the ability of new regional or national Earth stations to access different networks,
it quickly emerged that most Earth-station satellite dishes would in fact be
relatively easy to reorient toward a rival space segment, a feature known as
"steerability." In the 1960s, this feature of Earth stations raised, for the US
and Intelsat, the specter of insecure alliances and changing loyalties, particularly among the countries of the Global South.

As new Intelsat-certified Earth stations began to open around the globe
in the first half of the 1970s, the standard rhetoric that accompanied their
opening emphasized the power of communications to bring people together.
Talking points for a congratulatory call from US president Richard Nixon to
King Hussein of Morocco, on the occasion of the opening of the first Earth
station dedicated to communications satellites on the African continent
on January 7, 1970, emphasized the longstanding diplomatic connections
between the two countries, reaching back to an exchange of letters between
George Washington and Emperor Mohammed III in 1789.[46] The new Earth
station would, Nixon was to tell King Hussein, "draw people of United States
and Morocco closer together in the new decade."[47] As in the case of Yugoslavia's Intelsat Earth station, the opening of the Moroccan Intelsat Earth station was depicted as creating a firm and enduring alliance between Morocco
and the US.

But how strong *was* the connection that a new Earth station forged
between the country that owned it and the US, or even the Intelsat network?
As US officials quickly realized, even after constructing an Intelsat Earth station, developing countries could still reconsider their network membership
choices: an Intelsat Earth station's large antenna could be repositioned to
point at a different satellite in the sky. This was an essential technical feature
for several reasons. First, Intelsat was launching new satellites regularly in the
late 1960s and 1970s, and Earth stations needed to be adjustable to receive
signals from new satellites in new orbital positions. Conversely, when a new
satellite failed, Earth stations had to be able to adjust their position rapidly
to receive traffic rerouted to backup satellites on short notice. Furthermore,
even if there was no need to reposition the satellite dish to a new satellite,
some amount of positional drift still occurred, requiring adjustments on the
ground.

As a result, all large and small Earth stations had the potential to be reoriented toward a new satellite, potentially from another network. Just days after the Intersputnik announcement in Vienna in August 1968, J. D. O'Connell, a White House official, wrote to Anthony Solomon, assistant secretary of state for economic affairs, to urge him to consider having Intelsat provide the space segment for US domestic satellite services as a way to undermine calls for separate, regional satellite systems around the globe by demonstrating the US commitment to a single system. As O'Connell pointed out, even if countries had already invested in an Intelsat Earth station, there would be "no insuperable problem in reorienting the earth station antenna toward a different space segment," such as one provided by the Soviet Union or another regional network. This was a political as well as an economic threat. O'Connell understood the poverty of developing countries and the relatively low cost of building an Earth station as sources of political unreliability. "Even though there are sixty-two nations in INTELSAT," O'Connell warned, "more than forty are 'developing countries'—some of which might be persuaded by the Soviet Union or others to abandon INTELSAT, because their investment and commitment are not that great." He urged in conclusion, "I am sure that you will agree that the United States ought to make every reasonable effort to keep the faces of these nations turned toward the West."[48]

Moreover, the relative ease with which Earth-station operators could potentially access both Intelsat and Intersputnik space segments created additional concerns. The fear that developing countries might simply change satellite networks in response to lower costs was only the beginning. As Abbott Washburn warned Leonard Marks, the chairman of the US delegation to the Intelsat negotiations in the fall of 1969, "There is the possibility of earth station operators who are not members of INTELSAT 'poaching' on signals transmitted over the INTELSAT system."[49] Similar concerns about uncontrolled access to satellite signals had been raised by US officials since the passage of the original Communications Satellite Act of 1962: an illustration of these fears appeared in a 1962 report by the Congressional Legislative Research Service, depicting a satellite launched by country A, the signal from which was traveling down to an Earth station in a country "*unfriendly* to country A"[50] (see figure 4.8). This "unfriendly" country, moreover, was depicted as being located across a large wall, stretching off into the horizon, a clear reference to divided Europe a year after the erection of the Berlin Wall.

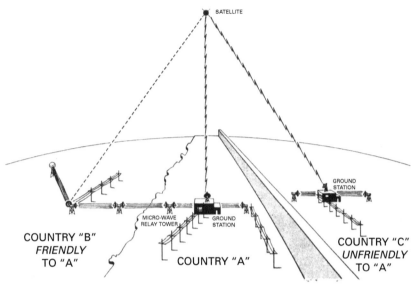

Political vs. Technical considerations in ground station location

FIGURE 4.8
Signal poaching diagram created by the Legislative Reference Service of the Library of Congress.

This particular threat marked an ambivalence at the heart of Intelsat's expansion. On the one hand, Washburn and his colleagues in 1969 were not overly concerned about signal poaching by "unfriendly" Earth stations. Washburn told Marks that while poaching "seems to be an area of real concern for the broadcasters," he felt that defending the broadcasters' intellectual property rights was beyond the scope of Intelsat's responsibilities.[51] Moreover, as we saw in chapter 3, US officials, like their Soviet counterparts, had long assumed that even if the Soviet Union and its allies did not join Intelsat, the Soviet network, whether domestic or international, would nonetheless be integrated with Intelsat's network outside the context of formal membership. Still, the fear of "unfriendly" Earth-station operators accessing Intelsat signals inverted the fantasy of satellite footprints as equal and undifferentiated across the territory that they covered. Not only was ground infrastructure necessary to realize the promise of satellite coverage zones, but the nature of signal distribution within satellite footprints, thanks to steerable antennae and signal poaching, was also potentially unstable, contested, or unwanted.

REPRESENTING EARTH STATIONS

Beyond the relatively narrow circle of technocrats, aerospace corporate reps, and state telecom officials engaged by Intelsat's Earth-station seminars, the new networks of satellite Earth stations were publicized to multiple audiences by national telecom administrations and states. Like satellites themselves, Earth stations were both connected to previous media networks and unprecedented. Unlike "media houses"—radio and then television broadcast centers—and computer network hubs that were based in urban centers, pre– direct broadcast satellite Earth stations were always sited in remote, rural locations to minimize radio signal interference with these other network hubs.[52] Since they were remote, like satellites themselves, Earth stations had to be explained, celebrated, and generally made visible as part of the promotion and sale of satellite communications to the state telecom officials and, ultimately, global residents who were asked to support investment in the construction of an expensive and not immediately fiscally self-sustaining medium. These new global infrastructures thus had to be presented in ways that would connect investment in an Earth station to stories of national achievement, modern infrastructural power, and access to space itself.[53] In the 1960s and 1970s, telecommunications officials and governments thus produced a large number of images and promotional materials featuring satellite Earth stations. These images sought to reassure their audiences by presenting space as both apolitical and discretely national to avoid fears about the fluidity of Cold War alliances or uncontrolled cross-border media flows.

One especially widespread set of images designed to acquaint the public with this new medium and its infrastructures were those nested within another global communications system: postage stamps, issued by many countries in celebration or commemoration of the opening of an Earth station in their countries or their entry into either Intelsat or Intersputnik.[54] Earth-station postage stamps, as a group, tended to represent space as both apolitical and discretely national, rather than global, transnational, and contested. They suggest an effort to deemphasize, if not conceal, the global and globalizing nature of this new technology in favor of satellite Earth stations as chiefly or exclusively a national technical achievement.

Many stamps commemorating the opening of an Earth station focus very narrowly on the station itself and its futuristic modern architecture, often set in a characteristic national landscape in which plant life and topography serve to anchor the otherwise generic space technological object.[55] This is true of the barren hills and lush tropical flora, respectively, of the Icelandic and Gabonaise stamps in figures 4.9a and 4.9b. Additional text on these two stamps connects them to national historical narratives as well. The 1981 Icelandic stamp celebrates seventy-five years of Iceland's telecommunications links with the outside world, connecting the 1980 opening of the Skyggnir Earth station to the opening of the first undersea cable connection from Iceland to Scotland in 1906 (see figure 4.9a). The Gabonese stamp (figure 4.9b), like many others, includes in its picture of the new Earth station another form of media infrastructure—a radio broadcast tower—making visible the connection between the Earth station and the rest of Gabon's media network. To further connect the new Earth station to a specifically national history, Gabon's stamp highlight's the new Earth station's name, which links the station to the date (December 2, 1967) when Gabon's president, Omar Bongo, took power. The only hint of the postcolonial politics of Earth-station construction is the artist's signature, visible at the bottom left: Rene Quillevic was a French engraver who made his name creating stamps for former French colonies. In both of these stamps, as in the many other national stamps that

FIGURE 4.9
Postage stamps depicting Earth stations: (a) Iceland (1981), (b) Gabon (1973), (c) Greece (1970), (d) Republic of Djibouti (1980), (e) Israel (1972).

feature only the Earth station itself or the Earth station with a radio tower, there is no sense of the transnational network or the Cold War alliance, with all the associated obligations and fears of influence and betrayal, to which the Earth station linked, however tenuously, each country that built one.

Some other countries' Earth-station stamps did, however, gesture toward the global nature of satellite communications and make the connection between the "ground segment" and "space segment" explicit by showing a satellite in space in visual dialogue with the new Earth station. Nonetheless, representations of the connections satellites made between *countries*, connections that might dangerously resemble previous colonial relationships, were quite rare. A Republic of Djibouti stamp follows the conventions already outlined, featuring a national landscape with desert sands and a palm tree, a radio tower in the background, and, on the upper-right side, a satellite (figure 4.9d). Although the satellite is there, it is unmoored from the geopolitical and institutional contexts that placed it in space. Instead, it appears as a purely technical fact, without reference to the kinds of international communications and relationships on Earth that communications satellite infrastructure facilitated.

The presentation of a decontextualized satellite is even more striking in an Israeli stamp, where even the bare technical representation of the satellite is softened and domesticated through an artistic representation that is rounded, colorful, and cute, if also somewhat psychedelic (figure 4.9e). Here, the "peaceful use of outer space" appears in the visual vocabulary of global hippiedom. A Greek stamp from 1970 (figure 4.9c) offers the most explicit reference to the satellite as a medium of global interconnection. Here, a satellite dish—perched upon what appears to be a Grecian column—points at a satellite hovering above two projections of the Northern Hemisphere, indirectly representing the transatlantic connections that the satellite makes possible, which link Greece, along with all of Europe and North Africa in the map on the right, to North and Central America (on the left).

Postal stamp images from Intersputnik member-countries, by contrast, reflected a greater comfort with the idea of global connection and influence—at least within the socialist world. These stamps were more likely to feature images of foreign flags and of the globe. Yet European socialist-bloc Earth-station stamps nonetheless share several key features with their Intelsat counterparts, including representations of a direct dialogue between a single country and a satellite in the sky above. A Cuban stamp, published to celebrate the opening of Cuba's Intersputnik station in 1973, depicts a Molniya satellite positioned directly above Cuban territory (figure 4.10a). Another stamp in the series, however, included images of Intersputnik member-country flags alongside a representation of signal beams traveling from the Molniya satellite down to Earth on both sides of the Atlantic, specifically visualizing Cuba's new connection to Eastern Europe via Intersputnik.

Other Intersputnik member-country stamps, from Soviet Union, Vietnam, and Czechoslovakia, all include the recognizable, circular Earth-station buildings of the Intersputnik network, but they notably also include humans and human activity. A 1981 stamp from the Soviet Union (figure 4.10b) portrays a group of people gathering in front of a large television screen outside a yurt, suggesting that they belong to one of the nomadic peoples of Central Asia. They are watching a broadcast of a technical object, perhaps a human-operated space vehicle, floating in space. Against the backdrop of a distant mountain range, an Intersputnik Earth station can be seen, with the antenna directed west, presumably toward Moscow. By including imagery of indigenous people in traditional clothing juxtaposed with Space Age television and satellite communications technology, this stamp offers an even more explicit version of the contrast, typical of US and European promotional images of satellite infrastructures as well, between the ultramodern space and broadcast infrastructures of satellite communications and the undeveloped landscapes in which they were set. A Vietnamese stamp from 1983 (figure 4.10c) similarly echoes some of the visual features of Intelsat stamps. This stamp depicts a person in front of an Intersputnik Earth station, operating a radio transmitter. The operator wears headphones and speaks into a microphone. Like Vladimir Nesterov's 1965 painting *The Earth is Listening,* this stamp emphasized the power of satellites to enable communication across distance. At the same time, this Vietnamese stamp also resembles the Israeli stamp in figure 4.9e, with its inclusion of a rainbow in the background, suggesting that the conversation made possible by satellites fosters not only the peaceful use of outer space, but also peaceful relations back on Earth.

The tendency to include human figures in socialist Earth-station images supports Svetlana Boym's argument that, unlike the empty American "outer space," the Soviet "cosmos" was imagined as a harmonious realm where "human or divine presence is made manifest."[56] Nonetheless, that human imagery could sometimes escape the narrow bounds of official propaganda. A Czechoslovak stamp, for example, featured, floating above an

FIGURE 4.10
Postage stamps: (a) Cuba (1974), (b) Soviet Union (1981), (c) Vietnam (1983), (d) Czechoslovakia (1974).

Intersputnik Earth station, the celestial figure of a woman, drawn in the manner of the émigré Czech artist Alphonse Mucha, holding up a Molniya satellite (figure 4.10d).

It would be easy to assume that the chiefly national orientation of these stamps is simply a product of their genre. Stamps, of course, are produced by national governments to represent and circulate images of their achievements to both national and global audiences.[57] In this framework, the national focus of their imagery, which disconnects satellite infrastructures from their global networks, is not surprising and likely tells us less about satellites and more about stamps as a medium in general. But there are two important counterarguments.

First, despite the apparent differences between some Intersputnik stamps and Intelsat stamps, the assertion that socialist satellites existed to serve human needs on Earth performed a similar function—that is, it assuaged concerns about the contested, threateningly globalizing features of this new

FIGURE 4.11
Earth station, cover image of "The World's Earth Stations for Satellite Communications," COMSAT, December 1970.

infrastructural network. Much like the Intelsat member-countries' stamps, the Intersputnik countries' stamps offered a benign, apolitical account of technical modernization—satellite communications with a human face.

Second, these philatelic images closely resemble images of satellite infrastructure produced for the public by Western aerospace corporations. Images of Earth stations promoted by COMSAT, RCA, and other firms involved in Earth-station construction tended to highlight their connections to outer space, bringing the Space Age down to Earth. Much like the Telstar brochure illustration discussed in this chapter, a photograph of an Earth station featured on the cover of a COMSAT brochure (figure 4.11) uses a darkened sky in the background to link the Earth station to the starry sky and an imagined extraterrestrial landscape, beyond the reach of our Sun.[58]

Moreover, Intelsat promotional materials designed for audiences in the Global South also tended to include images of local technical experts working on satellite communications equipment, highlighting the way that Earth-station ownership built local technical expertise (figure 4.12a). The back cover of a 1970 issue of the US Information Agency's magazine *Topic* portrays a young Moroccan engineer, Mohamed Senhaji, in front of the 100-foot antenna of the satellite Earth station outside Rabat (figure 4.12b). The caption explains that he is one of many engineers around the world that helps "bridge the communications gap between nations," and Morocco soon will be able to communicate with North and South America, Europe, and Africa using satellites. The feature article inside the magazine explains how Intelsat and modern communication has brought "the promise of the 21st century to mankind, three decades early," and that the "nations of Africa and the Arab World are 'illuminated' by microwave beams from both Atlantic and Indian Ocean satellites."[59] Expanding on the Nixon administration's rhetoric about the significance of the Moroccan Intelsat station for US-Moroccan relations, the article emphasized how Intelsat membership was fulfilling the promise of infrastructure, connecting Morocco globally while also ostensibly transforming a whole cast of workers into modern technical professionals engaged in everything from burying coaxial cables to link the Earth stations to terrestrial networks to "analyzing test patterns received from high flying satellite."

Intersputnik's own promotional materials are somewhat difficult to find and access; the organization struggled with inadequate budgetary allocations for publicity, even as late as 1977, when Intersputnik representatives were participating actively at international Astronomical Congresses and other industry gatherings.[60] Among the photos preserved in the Intersputnik organization's archives and published in a corporate history of Intersputnik however, is an image of the Intersputnik Earth station in Mongolia that shares similar tropes with other Soviet depictions of space technology alongside indigenous people in traditional dress. Like the Minenkov illustration of a reindeer herder in the foreground of a map of Molniya Earth stations, this photo emphasized the promise of imperial modernization by positioning Mongolian horsemen in traditional dress in front of a Space Age Earth station (figure 4.13).[61] As in the Molniya illustration, traditional modes of communication and transport, in the form of reindeer and horses (long part of the highly effective overland postal networks established in what is now

FIGURE 4.12
Left, technicians at the Earth station in Sehoul, Morocco. Right, Mohamad Senhaji in front of the Earth station. *Topic Magazine*, No. 52, 1970.

Russia by the Mongol empire), were on display in juxtaposition to the Space Age satellite infrastructures.[62]

There is a final view of Intelsat's Earth stations—one that was purely for internal or US government consumption, in which Earth stations were depicted as nodes in a highly centralized network. In this 1983 image, Earth stations appear as mere place names, arrayed on spokes running out from the satellite that serves them (figure 4.14). One would, of course, not expect governments that had lately acquired a new Earth station to create images like this one, which visually subordinated each new station to an enormous Intelsat satellite occupying center stage. This is an imperial view of the Earth-station network, in which an Intelsat satellite sits securely in the center among a long list of network hubs, neatly grouped by continent and including several socialist countries by 1983, including both Romania and the Soviet Union. Compared to the many Earth-station postage stamps, this

FIGURE 4.13
The Intersputnik Earth station in Naran, Mongolia. Reproduced with permission,
Intersputnik IOSC.

graphic representation of the Intelsat Earth-station network tells a dramatically different story about *to whom* or *to what* Intelsat Earth stations connected
the countries in which they were built, as well as about the nature of that
relationship. Here, the network appears genuinely global, but also hierarchical and stable, rather than contested and unstable. Ironically, this image was
produced as part of Intelsat's efforts to defend itself against the first efforts,
by President Ronald Reagan's administration, to break Intelsat's monopoly
on US international satellite communications. This image of a centralized,
stratified satellite communications network, never an accurate representation even at the height of Intelsat's dominance, would soon be completely
unimaginable.

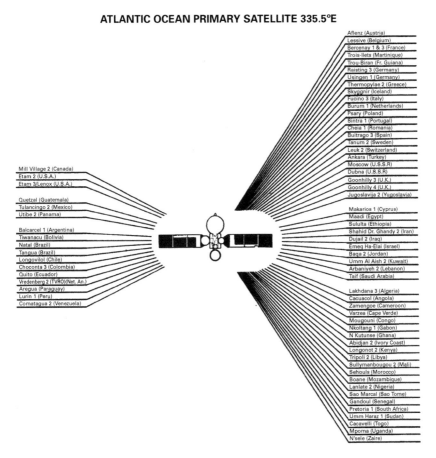

ATLANTIC OCEAN PRIMARY SATELLITE 335.5°E

Aflenz (Austria)
Lessive (Belgium)
Bercenay 1 & 3 (France)
Trois-Ilets (Martinique)
Trou-Biran (Fr. Guiana)
Raisting 3 (Germany)
Usingen 1 (Germany)
Thermopylae 2 (Greece)
Skyggnir (Iceland)
Fucino 3 (Italy)
Burum 1 (Netherlands)
Psary (Poland)
Sintra 1 (Portugal)
Cheia 1 (Romania)
Buitrago 3 (Spain)
Tanum 2 (Sweden)
Leuk 2 (Switzerland)
Ankara (Turkey)
Moscow (U.S.S.R)
Dubna (U.S.S.R)
Goonhilly 3 (U.K.)
Goonhilly 4 (U.K.)
Jugoslavija 2 (Yugoslavia)

Mill Village 2 (Canada)
Etam 2 (U.S.A.)
Etam 3/Lenox (U.S.A.)

Quetzal (Guatemala)
Tulancingo 2 (Mexico)
Utibe 2 (Panama)

Makarios 1 (Cyprus)
Maadi (Egypt)
Sululta (Ethiopia)
Shahid Dr. Ghandy 2 (Iran)
Dujail 2 (Iraq)
Emeq Ha-Elai (Israel)
Baqa 2 (Jordan)
Umm Al Aish 2 (Kuwait)
Arbaniyeh 2 (Lebanon)
Taif (Saudi Arabia)

Balcarcel 1 (Argentina)
Tiwanacu (Bolivia)
Natal (Brazil)
Tangua (Brazil)
Longovilol (Chile)
Choconta 3 (Colombia)
Quito (Ecuador)
Vredenberg 2 (TVRO)(Net. An.)
Aregua (Paraguay)
Lurin 1 (Peru)
Comatagua 2 (Venezuela)

Lakhdana 3 (Algeria)
Cacuacol (Angola)
Zamengoe (Cameroon)
Varzea (Cape Verde)
Mougouni (Congo)
Nkoltang 1 (Gabon)
N Kutunse (Ghana)
Abidjan 2 (Ivory Coast)
Longonot 2 (Kenya)
Tripoli 2 (Libya)
Sullymanbougou 2 (Mali)
Sehouls (Morocco)
Boane (Mozambique)
Lanlate 2 (Nigeria)
Sao Marcal (Sao Tome)
Gandoul (Senegal)
Pretoria 1 (South Africa)
Umm Haraz 1 (Sudan)
Cacavelli (Togo)
Mpoma (Uganda)
N'sele (Zaire)

FIGURE 4.14
Graphic from Richard Colino, "INTELSAT: The Right Stuff," manuscript, 1983. The Eisenhower Presidential Library, Abilene, KS, Abbott Washburn Papers, box 215, folder "Orion Challenge 1983 (1)."

All these images—produced by and for different participants in the global expansion of satellite communications infrastructure—downplayed transnational media flows or globalizing commercial ties in favor of nationalizing stories of Space Age technical achievement and direct, mystical contact with space itself. The architecture of the Earth stations themselves and the photographs and drawings used to promote them presented them as futuristic places where earthly conflicts were either irrelevant or could be overcome: Moon colonies here at home. In this sense, satellites' terrestrial infrastructure offered a mirror image of what human spaceflight promised in the 1960s—rather than escaping from Earth, transforming it in space's image. Public-facing and even internal representations of communications satellite ground infrastructure thus worked to assuage what were in fact substantial anxieties about the effectiveness and stability of both ground infrastructure itself and the political and economic relationships that underpinned it.

Considering these representations of Earth stations in the context of the contested, transnational history of satellite communications infrastructure in this period reveals how claims about US dominance in satellite communications and the idea that the US and Intelsat built a global satellite infrastructure in an orderly way, "country by country," was a political claim rather than a description of reality. Actors occupying different positions within this network, including corporations building Earth stations, national governments constructing them, Intelsat's leadership, and others—drew on Cold War geopolitical, economic, and cultural discourses to represent Earth stations in ways that reflected their particular symbolic and political agendas. Alongside US rhetoric linking the free flow of global communications to world peace, these images of satellite infrastructure worked to make a potentially threatening, expensive, globalizing, and regionalizing new media network more palatable to diverse global audiences, from national telecom officials to the broader public. At the same time, these images also highlight the many different, frequently conflicting understandings of what opening a satellite Earth station really meant for the future.

CONCLUSION

The symmetrical oval of a satellite's footprint on a world map concealed a great deal: the need for Earth stations that could distribute the signal and the limits of those distribution networks relative to the satellite's potential reach, the complex, regional, and economic negotiations over Earth-station site selection, and the labor and diplomacy needed to maintain a network of Earth stations. The instability of the Cold War relationships were materialized in the process of Earth-station sales, construction, and celebration, throughout which US government, corporate, and Intelsat actors were influenced by the asymmetrical but real threat of a rival, Soviet-led network.

Examining the history of communications satellite ground infrastructure and the dialogue between ground and space segments offers a more nuanced

understanding of the emergence of contemporary satellite systems. Unlike
the strictly bilateral, station-by-station expansion of Intelsat described by
Washburn in 1969, or the purely national story of access to Space Age modernity via the construction of a new Earth station, the construction of satellite
ground infrastructure was shaped by a great deal of transnational interaction, influence, and integration across geopolitical boundaries. With the construction of satellite Earth stations, this integration took a considerably more
material form, allowing a rethinking of global interconnectedness.

Attending to these aspects of the expansion of satellite ground infrastructure allows us to make visible a past in which the early days of satellite communications' global expansion looked far more like the current,
highly globalized, and fragmented satellite communications sector. This past
includes, moreover, a more significant role for the socialist world in shaping satellite infrastructures as both an imagined threat and as a surprisingly
eager partner, whose own vision of the future of satellite communications,
reflected in both the actions and representations of new Earth stations, was
not so very different from that of Intelsat and its member-countries.

5   HOTLINES, HANDSHAKES, AND SATELLITE EARTH STATIONS: INFRASTRUCTURAL GLOBALIZATION AND COLD WAR HIGH POLITICS

We met Dr. František Šebek at an Italian restaurant in Prague's old town. Dr. Šebek was the former technical supervisor of the Sedlec-Prčice communications satellite Earth station in the Czech Republic.[1] The Earth station that Dr. Šebek ran was built in 1974 outside Sedlec-Prčice, a small town about forty minutes west of Prague (figure 5.1). From 1974 until 1989, it was a Soviet Intersputnik Earth station, an Orbita-2 model transmitting television programming and telephone calls among the network's socialist-bloc neighbors. The Earth station's local staff, many of whom were women, lived in the village, while supervisors like Dr. Šebek lived in Prague and traveled there mostly to help repair things when they broke down. The local staff had a hunting club and enjoyed hunting in the forest around the Earth station during their leisure time, when they were not coordinating the flow of television programs and international telephone connections or watching an entertaining Cuban television show that Dr. Šebek remembered fondly. Beginning in 1989, however, the Earth station's function expanded dramatically: starting that year, it featured three large, steerable satellite dishes. One pointed, as before, at the Soviet Union's geostationary Statsionar satellite over the Atlantic Ocean. Two other, newly built antennae pointed at Intelsat satellites over the Indian and Atlantic oceans, sending and receiving signals from both.

We expected the memory of this change in Sedlec-Prčice's status, from an exclusively socialist-world Earth station to one integrated into Intelsat's much larger global network, to have been quite sharp in Dr. Šebek's memory. But in fact, Dr. Šebek's stories of his work at the Earth station, both before

FIGURE 5.1
The Intersputnik Earth station in Sedlec, Czechoslovakia. Reproduced with permission, Intersputnik IOSC.

and after 1989, were equally global. Dr. Šebek explained that as a child, his imagination had been captured by space exploration, but as someone growing up on the Soviet Union's imperial periphery in Czechoslovakia, he knew that he could not become an astronaut. Satellite communication, by contrast, with its widely distributed ground infrastructures—far more numerous than launch sites around the globe—allowed him to build a career in space technology. Dr. Šebek's work at an Intersputnik station allowed him to travel internationally as well. While most of the engineers who traveled to set up new Intersputnik Earth stations were Soviet citizens, Dr. Šebek was able to travel occasionally. With some delight, he shared an anecdote in which, on a trip to install new equipment in Cuba's Intersputnik Earth station (figure 5.2), he tricked his less experienced Russian counterparts by asking them to point out, in the night sky, a black hole whose radio emissions they had used to test and calibrate the new Earth station's signal reception. Perhaps most important, Dr. Šebek and his coworkers had already participated in global flows of information, including from Intelsat member-countries, before Sedlec-Prčice itself became an Intelsat station. That was because Sedlec-Prčice, in 1989, was far from the first Earth station to draw signals from both Intelsat and

FIGURE 5.2
The Intersputnik Earth station in Caribe, Cuba. Reproduced with permission, Intersputnik IOSC.

Intersputnik satellites. From 1974, the Soviet Union had owned and operated three Intelsat Earth stations—one in Dubna outside Moscow, one in Vladimir, and one in Lviv, in western Ukraine (figure 5.3).[2] Moreover, several other Intersputnik member-countries already operated both systems, including Cuba (since 1979) and Nicaragua (since 1986).

The story of how the Soviet Union and several Intersputnik member-countries came to be integrated into Intelsat as Earth-station operators offers a fresh look at the relationship between infrastructural globalization and Cold War high politics. Much recent work on trans–Iron Curtain scientific exchange and infrastructural integration has focused on the ways in which both these activities proceeded largely undisturbed by military and diplomatic events at the highest level. An example of the relative insignificance of fluctuations in Cold War politics for these spheres is the nearly nonexistent impact that the Soviet invasion of Czechoslovakia had on negotiations over the possibility of Soviet entry into Intelsat. Sari Autio-Sarasmo has observed a similar lack of response to the invasion in the realm of scientific and

FIGURE 5.3
Google Earth image of the Lviv Earth station, 2022.

technical exchanges between Eastern and Western European countries.[3] Per Högselius likewise argues that the construction of natural gas pipelines that linked Eastern and Western Europe proceeded largely without regard to prevailing geopolitical divisions or events.[4] This view of the essential disconnect between infrastructural projects and scientific exchanges, however, tends to risk reinforcing the claims of Western government officials involved in scientific, technical, and media exchanges that these activities were genuinely disconnected from Cold War diplomacy. As Audra Wolf demonstrates, scientific exchanges were highly politicized activities that directly promoted state goals during the Cold War, even if they functioned differently from, or even contradicted, high-level diplomatic interactions.[5] Because the development of satellite communications infrastructures was always linked to events in the higher-profile world of human spaceflight, we argue, those infrastructures were in fact directly shaped by Cold War high politics, though not always in the ways that we might expect.

Indeed, the integration of the Soviet Union and several other Intersputnik members into Intelsat as Earth-station operators was the result of decisions made at one peak of Cold War high politics: the Nixon–Brezhnev summit of 1972. The 1972 summit meeting culminated, most famously, in the highly symbolic Apollo–Soyuz joint docking mission of 1975. Less noticed, and far from the main event at the summit, was the decision to create a satellite-based Moscow–Washington hotline as a backup to the existing cable hotline. While this hotline was ultimately a more reliable version of its cable predecessor, its construction required permanently linking Intelsat and the Soviet Molniya networks via the exchange of Earth stations capable of receiving from and transmitting to the other networks' satellites. This apparent footnote to the Apollo–Soyuz test project and the Strategic Arms Limitations Talks (SALT) negotiations brought about an integration of equipment and networks across the Iron Curtain that was far more permanent than the famous handshake in space between astronauts and cosmonauts. It also had a significant impact on media flows across the Iron Curtain and facilitated the de facto integration of the Soviet Union into Intelsat, fifteen years before it officially joined the network in 1991. The process by which satellite communications networks became globally integrated infrastructures, facilitating increasingly routinized global media flows and human and institutional interactions, was in part an unintended consequence of Cold War nuclear summitry. Moreover, as the unintended consequences of the exchange of Earth stations from each network became apparent, it was the supposedly closed and autarkic Soviet Union and other socialist states that increasingly embraced transborder information flows and interactions, while the US worked to prevent them.[6]

AN EXCHANGE OF EARTH STATIONS

The first summit between Richard Nixon and Leonid Brezhnev in May 1972 is remembered chiefly for the SALT agreement, negotiations regarding the Vietnam War and other Cold War hot spots, and the signing of agreements that would result in the Apollo–Soyuz test project.[7] Like the pursuit of international cooperation in space under Nixon more generally, the summit was part of Nixon and Secretary of State Henry Kissinger's search for ways to respond to domestic unrest and the antiwar movement.[8] While the Apollo–Soyuz test project already had been in planning for a couple of years, the summit provided an opportunity to put in motion the symbolic power of space. Nixon actively promoted the idea of "space brotherhood," building on an image of astronauts and cosmonauts shaking hands in space to counter the risk of imminent nuclear war.[9] But the summit had another outcome as well: the creation of a satellite backup to the original, 1963 cable Direct Communications Link (DCL) implemented in the wake of the Cuban Missile

Crisis to prevent unintentional nuclear disaster by improving real-time communication between the two Cold War rivals. Unlike the existing cable hotline, which was constantly vulnerable to damage from, to give two real-life
examples, a Finnish farmer plowing and a manhole fire in Baltimore, the
new hotline allowed direct communication via space.[10]

The negotiations to improve the direct communications link (DCL)
between Washington and Moscow, also known as "the hotline," by creating a second communications satellite linkup occupied a very lowly place in
the hierarchy of agreements negotiated in advance of the May 1972 Nixon–
Brezhnev summit. The countless memoranda and letters exchanged between
Kissinger and Anatoly Dobrynin, Soviet ambassador to the US, as well as
their regular in-person meetings in the lead-up to the summit, constituted
a relationship known as "The Channel"—a name that suggests how much
of détente negotiations depended on a combination of personalism and
the traditional media of diplomacy, diplomatic cables, and in-person meetings rather than high-tech satellite link-ups that were definitively *not* "The
Channel."[11] These conversations addressed issues of concern between the
two sides in a standard order, with active military conflicts such as Vietnam
or India–Pakistan leading, followed by issues in the SALT negotiations, then
trade agreements, and (always last) a brief mention of scientific cooperation,
negotiations that were purposefully conducted by lower-ranking diplomats
and scientists. The construction of the satellite hotline did not fall even into
this latter category since it was a national security infrastructure project that
was explicitly excluded, for reasons described next, from use in future commercial or scientific exchanges.

Like scientific cooperation, however, the hotline project could serve as a
way for diplomats from both sides to pursue larger political goals outside
the public gaze. The sole mention of the satellite hotline in the Kissinger–
Dobrynin exchanges in 1971–1972 came in a May 1971 conversation between
Dobrynin and Kissinger in which Kissinger proposed that the real purpose
of a visit to the White House by Soviet deputy minister V. S. Semenov—the
discussion of restrictions on antiballistic missile systems—could be concealed by inviting Semenov to the White House to "sign an agreement or an
arrangement regarding an improvement of the direct communications link"
via the construction of a satellite hotline, which had been negotiated earlier
and could now be formalized.[12] From Kissinger's perspective, the satellite
DCL was a minor technical sideline to the summit, chiefly useful as a ruse to

conceal the real work of negotiations. Nonetheless, this infrastructural sideline to high politics at the SALT talks had the effect of integrating the Soviet Union into Intelsat.[13]

Planning for the construction of the Molniya Earth station in the US and two Intelsat stations in the Soviet Union began after agreement on the new satellite hotline was reached in September 1971. On Earth, these airy radio signals would be sent and received by heavy, costly ground stations that needed to be built. On the US side, this meant constructing a Soviet Molniya station at Fort Detrick in Maryland and the reorientation of a COMSAT Earth station in Etam, West Virginia, to receive signals from Soviet satellites. This meant the construction of an initial two Intelsat stations, followed shortly by a third, in the Soviet Union.[14]

Together, these new Earth stations created two sets of links—one via the Intelsat IV satellite over the Atlantic and another via four Soviet Molniya II satellites in highly elliptical orbits (later, the US Molniya Earth stations pointed to a geostationary Soviet Gorizont satellite). Each of these circuits was connected by at least two nuclear-attack-proof, redundant cable and microwave circuits to terminals in the Pentagon and the White House Situation Room, respectively, as well as the Soviet Ministry of Defense, the Kremlin, and Communist Party headquarters.[15]

Despite the focus on nuclear "survivability" for the new DCL's ground infrastructure and cable connections from the Earth station to political and military command centers in each country, the DCL's uses and impact were neither inherently limited to national security purposes nor even especially well suited to them.[16] Certainly, the new DCL, which came into operation in 1978, fulfilled hopes for a fully reliable communications link, not subject to unexpected disruptions that might imperil the globe.[17] However, traffic on this new, reliable satellite hotline link also was potentially vulnerable to interception from anywhere within its footprint on the ground. Within this footprint, any Intelsat "or other comparable ground stations operated by other interested parties" could potentially intercept, identify, and demodulate any communications channel using easily concealed antennae.[18] While these transmissions could be encrypted, the new DCL became part of much larger, multidirectional information flows that mingled commercial, intelligence, and defense uses.

Beyond its primary utility as a more reliable communications link between superpowers, the satellite DCL project's most meaningful outcome was infrastructural integration. The new Earth stations on each side made political détente material, integrating space infrastructural networks that, while always relatively technically compatible and mutually constituted, had been largely discrete. Construction, on both sides, took place in partnership with commercial satellite technology firms, launching what became extensive relationships and interactions between the Soviet Union and capitalist-world aerospace corporations. On the US side, Harris Corporation constructed the Molniya-compatible Earth station in Fort Detrick, together with the US Army Satellite Communications Agency, while the Soviet Union Ministry of Communications built Intelsat Earth stations outside Moscow and Lviv with technical assistance from ITT Space Communications, a subsidiary of ITT World Communications (the firm that helped build the original cable hotline).[19] These contracts entailed ongoing training and orientation of Soviet personnel by ITT staff which, as the firm boasted in its newsletter, "will give engineering and technical personnel an opportunity to work together for the future."[20] These contacts were part of dramatic expansion of Soviet meetings with and purchases from global satellite communications firms from the early 1970s onward, including with the Japanese telecommunications firm NES.[21] Given the expansion of Soviet relationships with multinational aerospace firms from the 1970s onward, the construction of ground infrastructure for the new DCL link facilitated not only a limited technical integration of Soviet satellites into global networks, but also a broader economic integration into global capitalism, which Soviet decision-makers actively sought in the 1970s.[22]

Indeed, the possibility that a Soviet Molniya Earth station in the US could be used for commercial purposes was what most concerned the US officials in the Defense Department charged with planning the DCL in the summer and fall of 1971. In a November 30, 1971, memo to Kissinger outlining plans for implementation of the new DCL agreement, David Solomon noted the legal and political problems posed by the presence of a Molniya Earth station on US soil.[23] The Molniya II Earth station had to be acquired by the Department of Defense because ownership or operation by a US telecommunications company would "carry the connotation of a commercial telecommunications service."[24] While "it is conceivable that the USSR may in the future wish to use the MOLNIYA II system for commercial telecommunications services between their country and the US, as well as to other countries," the memo continued, "it would not be appropriate for the MOLNIYA II system to be used by the US for any purpose other than the satellite DCL in view of the US commitment to INTELSAT."[25]

Moreover, in the classified annex to Solomon's memo, he argued that the Molniya station should be designated as exclusively for national security purposes for two reasons. The first, in the short term, was to prevent a situation in which the Intelsat Board of Governors was in a position to vote up or down on an agreement already signed by the US government. And the second, in the longer term, was to prevent the Soviet Union from using the US-based Molniya Earth station for commercial purposes. "If the Molniya II earth station is owned by the US Government and located on US Government property," Solomon pointed out, "then the US is in a strong position to decline possible USSR requests for use of the Molniya II system for commercial telecommunication purposes on the basis that this use is contrary to US Government regulations."[26] The decision to acquire the Molniya station through the Department of Defense and locate it on a US military base at Fort Detrick was motivated not only by the need to meet the legal requirements to preserve COMSAT's monopoly on satellite communications originating in the US, but more broadly to block Soviet commercial activity in satellite communications on US soil.

While these efforts were narrowly successful in limiting the commercial use of this specific Molniya Earth station at Fort Detrick, they did nothing to prevent the substantial integration of the Soviet Union into Intelsat's infrastructural and institutional networks, just as they did not prevent the Soviet Union from entering contracts with international space telecommunications firms. The construction of these Earth stations meant that, for the first time, Soviet technical personnel were able to work directly with Intelsat's equipment and technical protocols.[27] After the Soviet Union's Intelsat stations in Dubna, Vladimir, and Lviv were put forward for and successfully passed a review confirming that they met Intelsat's technical standards, the Soviet Union became an official Intelsat Earth-station operator.[28] As COMSAT's Geneva office reported in a November 1971 letter to the Foreign Relations Department in the Ministry of Communications, Soviet representatives had inquired, during a meeting in Venice (rather immediately, given that the DCL agreement was signed in late September 1971), about whether a nonmember-state, which nonetheless operated an official Intelsat station, would have the right to attend annual Intelsat conferences for all Earth-station operators. The Intelsat official could now confirm, he wrote, that "once the Soviet

INTELSAT station is approaching operational status, representatives of your ministry will be welcome at the Atlantic region Earth Station Operators Meetings," which took place roughly annually.[29] These meetings were an essential part of maintaining the functionality of Intelsat's global network of Earth stations, and operating an Intelsat Earth station ultimately required participation in these technical seminars, whether or not the possibilities for technology and expertise transfer to the Soviet Union were desirable from the US's perspective.

MAINTAINING SATELLITE NETWORKS: INFRASTRUCTURAL UPKEEP

Infrastructures by definition require ongoing maintenance and care to function and reproduce themselves across time.[30] In the case of satellite communications infrastructures, the technical, economic, and political connections between an Earth station in a particular country and Intelsat or Intersputnik's satellites in space constituted an ongoing relationship that required maintenance and investment. Earth-station operators also had to continually invest in the staffing, upkeep, and technical updating of their station to maintain its connection with a space segment that changed frequently. Alongside the grandiose rhetoric that sometimes accompanied their opening, these significant ongoing maintenance obligations were another key feature of the Cold War and postcolonial technical relationships embodied in satellite communications Earth stations.

One source for uncovering these ongoing relationships between Intelsat and its Earth stations around the world—including those operated by the Soviet Union after 1974—are the proceedings of a series of large, multinational technical seminars organized in partnership with a variety of professional and trade groups throughout the 1970s, often hosted outside the US to deemphasize American dominance of Intelsat, which remained a political problem.[31] These seminars brought together representatives of countries with Intelsat Earth stations and corporations that manufactured Earth-station equipment to provide the latest technical information, to alert station operators to coming changes, and, likely, to facilitate sales of updated equipment to Intelsat member-countries by the corporate participants.

The agendas of Earth-station seminars in the late 1960s and early 1970s reflected the political tensions underlying the construction of a global infrastructural network across significant differences of national wealth and political perspective in the context of decolonization and activism by countries of the Global South. Seminars featured predominantly US- and UK-based experts offering predominantly developing-world audiences detailed instructions on the process of setting up and running new and existing Earth stations, with a focus on top-down instruction by experts and little formal time for dialogue. Topics ranged from new measures for correcting signal interference produced by rainfall to detailed instructions on how to assemble and train technical staff.[32]

Unlike national security–related space infrastructure like tracking stations and launch sites located in countries of the Global South, Intelsat (and Intersputnik) Earth stations were owned by the telecommunications agencies of the countries in which they were located and staffed by citizens of those countries as well.[33] Perhaps as a result, these lectures sometimes reflected substantial anxiety on the part of Western experts about the process of obtaining and training qualified local staff to operate Earth stations in the Global South. Eager to transfer ideologies and procedures of information management as well as technology, Intelsat seminar organizers instructed Earth-station managers on how to manage information and procedures within the Earth station, down to an extraordinary level of detail. One such lecture even featured a photo of the ideal kind of small bookcase that each Earth station ought to include to ensure that employees had access to the relevant guidebooks and training manuals.[34]

The upgrading of Intelsat's space segment sometimes required changes that were difficult, expensive, or otherwise unwelcome for Earth-station operators. Seminar organizers moved to address participants' frustration by altering the format of these seminars to make them more dialogic. One such example, the planned introduction of INTELSAT-V satellites in the Atlantic and Indian oceans in 1979 and 1980, respectively, led the Intelsat Board of Governors to take a more active role in organizing an Earth-station technology seminar with a new format.[35] The technical features of the INTELSAT-V satellites necessitated substantial modifications to existing Earth stations, requiring them to begin employing dual-polarization for frequency reuse and the introduction of a new frequency band that INTELSAT-V satellites used. This meant that Intelsat Earth stations around the globe would have to have their antennas adjusted and retested to ensure that they would work with INTELSAT-V satellites.[36] In response to what seems to have been considerable concern among Earth-station owners about these changes, Intelsat's board officially endorsed and helped organize the 1976 Earth Station Technology Seminar in Munich, which also featured, for the first time, extensive time for questions from Earth-station owners' representatives about the kind and timing of technical changes required.[37]

Although Soviet participants were present at these seminars, the existence of Intersputnik rarely figured in the lectures and discussions at these seminars. This is perhaps not surprising since these seminars were designed to communicate and maintain Intelsat's technical standards for all new and existing Earth stations around the world, rather than address interactions with other networks. Nonetheless, the fact of ongoing interaction between the two networks does appear indirectly in the seminars' proceedings. Presentations regularly included references to non-Intelsat Earth stations, which Intelsat-network members used or with which they coordinated to provide coverage of events of special political or news significance. In the 1976 Earth-station seminar proceedings, as in other Intelsat reports from the mid-1970s, these stations were not identified explicitly as Intersputnik or Molniya Earth stations; instead, they were grouped, together with military and other non-Intelsat Earth stations in countries like France and Saudi Arabia, under the larger rubric of nonstandard Earth stations.[38] This language served both to erase Intersputnik's presence as an alternative to Intelsat—in the year when the Soviet Union began to offer geostationary satellite service via its Gorizont satellites—and to emphasize the shared technical standards and common infrastructures that connected Intelsat Earth-station owners to the larger network. In this sense, the invisibility of the Intersputnik network as an unspecified set of nonstandard Earth stations also constituted a form of infrastructural upkeep, both concealing and enabling ongoing interaction across Cold War geopolitical divides.

At the most basic level, the published proceedings of these Intelsat Earth-station operator seminars reveal how much technical information and interaction with other Intelsat operators became available to the Soviet Union once it became an Intelsat Earth-station operator from 1974, followed by Romania in 1977 and Cuba in 1979. If these socialist-world Intelsat stations had remained completely unused beyond their use for the DCL, they still would have served to connect Soviet officials and technical specialists to Intelsat's global human and institutional networks. But they did not go unused. The new Earth stations made possible expanded transmissions of telephone calls, telegraph and teletype messages, and television programming, leaving

Soviet communications officials and their counterparts in other Intersputnik member-countries with the question of exactly how the new Earth stations should be used.[39]

In early 1974, Soviet minister of communications N. V. Talyzin invited members of Intersputnik to participate in an "exchange of opinions about the use of the earth station for cosmic communications in the USSR (in the region of the city of Lvov [Lviv]) in the interests of the Intersputnik Organization's member-countries."[40] This conversation took place in Moscow in May 1974 and covered the assessment of member-countries' needs for Intelsat communications channels, as well as organizational and technical questions about how to use the station. Decisions were confirmed by the Ministry of Communications in 1974. By December 2, telecommunications officials had finalized plans for what would begin as five fixed telephone channels and two automatic assignment (SPADE system) telephone channels linking the Soviet Union and the US, with plans to eventually increase the Soviet–US fixed telephone channels to 15 and total fixed telephone channels to 60–100, "considering the growing needs of [other] socialist countries."[41] The December 1974 technical document also confirmed plans for US–Soviet television channels with audio as well as expanded telex and telegraph channels in the new Earth station. These were used actively by the Soviet State Television and Radio Broadcasting Committee from the late 1970s onward, including to receive televised reports filmed by the enormously popular and influential cohort of Soviet television journalists reporting from the US, and for the famous US–Soviet live televised satellite bridges of the late 1980s.[42]

BECOMING INFRASTRUCTURE: NICARAGUA'S TWO EARTH STATIONS

A 1984 Soviet report on the planned construction of an Intersputnik Earth station in Nicaragua suggests the extent to which satellite communications networks and infrastructures had become integrated across the Iron Curtain by the mid-1980s. In that report, experts from the Soviet Ministry of Communications evaluated plans for the new Intersputnik station, to be located outside Managua above the Laguna de Nejapa. The report concluded that the plans "deserved an outstanding grade."[43] What made this project outstanding, the report suggested, was not its quite modest capacity, which was limited to twenty-four phone channels (with the possibility of expansion to sixty in the future) and the ability to receive one black-and-white or color television channel (with the possibility to distribute television broadcasts from Nicaragua in the future).[44] Instead, what was remarkable about the new Earth station was its supremely efficient manufacturing and design—the whole Earth station was largely prefabricated and packed into just six shipping containers for easy and inexpensive assembly on site.

Perhaps the greatest evidence of this project's admirable cost efficiency, however, was its maximal reliance of the infrastructure already in place in the form of Nicaragua's existing Intelsat Earth station, opened in November 1972 under the US-aligned Anastasio Somoza dictatorship and located just 150 meters away on the same hillside. Since the Intelsat station was so close, the new Intersputnik Earth station could employ not only the existing station's electrical power source and backup generator, but, as the report noted with approval, also the personnel already working at the Intelsat station, and even the employee break room.[45] Accompanying the report was a hand-drawn diagram of the planned Earth station, depicting two identical Earth stations nestled on the same hillside, with arrows and an angle indicating their antennae's respective orientations toward the Intelsat and Intersputnik satellites over the Atlantic.[46]

In effect, the new Intersputnik Earth station in Managua treated the existing Intelsat station as infrastructure, following the pattern of infrastructural development that Lisa Parks and Nicole Starosielski describe as a "layering of an emergent system upon an existing one."[47] After the 1979 Sandinista revolution, in other words, Nicaragua did not end its relationship with Intelsat. Instead, it simply added another ground station and large satellite antenna—a Soviet TNA-77 like those that had been installed in Intersputnik Earth stations in Laos, Vietnam, and Afghanistan in the preceding years—in the very same spot. What were ostensibly separate Cold War satellite communications networks, corresponding to rival Cold War blocs, in fact were deeply entangled on the ground in hot spots like Managua, and were characterized as much by cooperation and coexistence as they were by competition, suspicion, and hostility.

A year before the opening of the Managua Intersputnik station, Intelsat's board of directors recognized the de facto situation of extensive interaction between Intelsat and Intersputnik member-countries and Earth-station operators, as well as the colocation, in multiple countries around the world, of Earth stations from both networks. In October 1985, the board of directors passed measures to equalize the status of Intelsat members and nonmember users

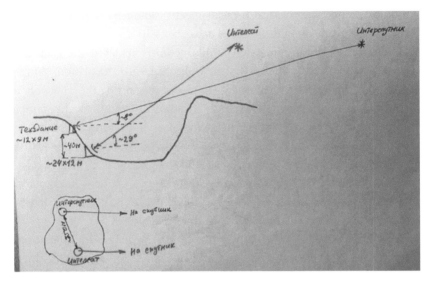

FIGURE 5.4
Hand-drawn diagram of the planned Intersputnik station in Nicaragua. The Intelsat
satellite and Earth station are in red; the Intersputnik ones are in green.

like the Soviet Union, as well as to approve requests from Nicaragua, Iraq, and
Algeria to use the Intersputnik system while retaining Intelsat membership.[48]

CONCLUSION

Both internal Soviet and US intelligence assessments of Soviet technical capacity in satellite communications in the 1970s emphasized the inferiority of
Soviet satellite communications technology, especially in its space segment.
Echoing Georgii Pashkov's 1973 assessment that "their satellite is much better
than ours," a 1976 Central Intelligence Agency (CIA) report on the Statsionar
geosynchronous communications satellites, the Soviet counterpart to the
INTELSAT-IV satellite, concluded that "the Soviets lag far behind INTELSAT in
communication satellite technology."[49]

These assessments failed to predict the significant extent to which the
Soviet Union was able to become integrated into Intelsat and expand Intersputnik globally, despite—indeed, arguably because of—its status as an asymmetrical actor, with far lower investment and less advanced technology, but
also lower prices and strong motivation to profit from its space technology.

Technical inferiority could also mean greater simplicity, ease of use, and lower costs. Thus, despite the lesser capacity of its communications satellites, the Soviet Union and Intersputnik members achieved the goals that they had articulated in the early years of Intersputnik's formation: gaining network members in the developing world and connecting their network to Intelsat. At the same time, this expansion was not a straightforward example of "hidden integration"; that is, integration of infrastructures a cross geopolitical boundaries without regard to political divisions. Instead, it was Cold War high politics itself—Nixon and Brezhnev's summitry and the decision to create a satellite DCL to back up the existing cable hotline—that provided the Soviet Union with its first Intelsat ground stations and first extensive cooperation with private multinational aerospace firms.

The impact of the new commercial contacts opened up by Intelsat participation was significant. From the mid-1970s onward, the Soviet Union actively pursued relationships with private firms, while also creating financial institutions to ease the process of investment within the socialist bloc. At the tenth annual Intersputnik board meeting in Brno, Czechoslovakia, in 1981, the Intersputnik member-country representatives voted to change Intersputnik's rules to allow it to borrow money from banks in countries where it operated, as well as from the International Investment Bank, a Soviet-bloc development bank created by COMECON (Council for Mutual Economic Assistance) members in 1970.[50] That same year, Intersputnik member-states asked the organization's directors to create a new price scale for Intersputnik's services, "taking into consideration the transition to the phase of commercial utilization of the communications system."[51] By the early 1980s, Soviet officials were taking fifteen to twenty meetings a month with multinational aerospace corporations.[52] In April 1984, to give just one example, the director and deputy director of Intersputnik (Soviet and Cuban citizens, respectively) met in Moscow with three Canadian businessmen to discuss leasing spare capacity on Intersputnik satellites to these Canadian partners, who would then resell it for use by "private, corporate, and regional networks." Payment, the Canadians promised, could be in goods as well as hard currency, "transferred via Bermudan banks."[53]

Media network integration also continued apace. Also in 1984, just twelve years after the Nixon administration sought to prevent commercial use of the US's new DCL Molniya Earth station by locating it on a military base at Fort Detrick, CNN founder and chief executive officer Ted Turner signed a long-term program exchange agreement with Intersputnik. Confirming US and Intelsat officials' original fears, in the mid-1960s, about the steerability of Earth stations, for an investment of only $10,000, Turner's CNN modified an Atlanta Earth station to receive Intersputnik transmissions. CNN initiated this relationship so it could cover the Friendship Games, the athletic competition hosted by the Soviet Union as an alternative to the 1984 Olympics in Los Angeles, which the Soviet Union was boycotting, at a fraction of the cost of covering the games using Intelsat.[54] But the games did not end this relationship; Turner gained permission from the Federal Communications Commission (FCC) to continue receiving Soviet news programming via Intersputnik.

From Intersputnik's perspective, the biggest problem was not negotiating deals with Western firms, but rather enforcing contracts with them. Intersputnik officials complained, in 1988, that "questions of CNN's payment" for its use of the network "had not yet been resolved." With Western partners failing to pay their bills, or paying them in barter or via Bermudan banks, the corrupt processes of privatization that we think of as beginning no earlier than 1988, with Mikhail Gorbachev's law on cooperatives, were well underway by at least 1984. Echoing the language of Intelsat's Earth-station operator training, in which Intersputnik Earth stations appeared as nonstandard, Intersputnik began including CNN's Atlanta Earth station on its list of Atlantic Ocean Earth stations under the rubric of "stations without official authorization."[55] In this context, we can see why the arrival of an Intelsat station in Sedlec-Prčice in 1989, was, while notable, not an entirely dramatic transformation. By the end of the 1980s, Intelsat and Intersputnik had already become significantly integrated and interconnected.

EPILOGUE

Tangled histories of global infrastructure and planetary imagination are hard to tell. For many years, a replica of the original Sputnik has occupied a place of pride just inside the entrance to Moscow's Museum of Cosmonautics, positioned as the first in a series of Soviet Space Race victories. As the museum's permanent exhibition continues, however, later Soviet satellites quickly disappear upward into remote corners of the main galleries in favor of the stories and objects of human spaceflight—hand-sewn space suits, spaceship capsules into which we can peer, and the cosmonaut Alexei Leonov's marvelous paintings of the human view from a spaceship window. Until 2022, another Sputnik replica greeted visitors to the National Air and Space Museum at the Smithsonian in Washington, DC, though its small size meant that it was a bit dwarfed by the *Spirit of St. Louis* and the lunar module from the Apollo 11 Moon landing, which hung nearby at that time. American and Soviet histories of the US–Soviet Space Race had thus long followed a similar path, narrating a timeline of national achievements from Sputnik to human spaceflight, framed in terms of a binary Cold War.

Recently, this has begun to change, thanks to the efforts of aerospace museum curators and a growing sense of the importance of the history of communications technologies within aerospace history, both among historians and industry representatives. Beginning in 2016, when the National Air and Space Museum's aging building required major renovations, its curators began planning a new gallery, "One World Connected," dedicated to the impact of aerospace technology on globalization, in which the history of application satellites, including communications satellites, plays a central role. The exhibit, which opened in October 2022, asks how we became globally connected, outlining the history and construction of infrastructures for aviation, undersea cable traffic, and artificial Earth satellites. Following Hannah Arendt's original observation that the view back at Earth from space marked a new era, "One World Connected" also translates, for a popular, youth audience, the extensive literature on the political and psychological impact of whole Earth photographs like "Earthrise." A central feature of the new gallery is an enormous, rather beautiful globe, onto which maps of human population, animal migration, and satellite internet coverage are projected in response to interactive touch screens (figure E.1). The globe positions viewers as astronauts, looking back at Earth, and surrounding exhibit panels encourage visitors to think of themselves as global citizens, connected by technology. (This perspective is made even more explicit with a walk-in model of the cupola on the International Space Station.) On one visit to the exhibit, we overheard an older man, looking at the globe with a young child in a stroller, explain that this is Earth, the big planet we all live on together. Lest we wonder who really does all this connecting, corporate sponsors are noted prominently throughout the exhibit. The gallery's central globe is sponsored by Iridium, a publicly traded US corporation that operates a communications satellite constellation.

Surrounding the interactive globe is a series of panels featuring large photographic portraits of individual people from around the world, holding objects that connect them globally, whether those are the communications technology that they use in their everyday lives or objects they have carried with them through their journeys as immigrants. This collection of work by multiple photographers from around the world is a successor to the numerous twentieth-century media projects that sought to represent global diversity in a single, global moment, including the two Soviet "One Day in the World" projects, as well as the photographer and Museum of Modern Art curator Edward Steichen's 1955 "The Family of Man" exhibit. Some of these portraits are powerful; others repeat the familiar trope of juxtaposing traditional dress and ways of life with modern technology. According to the gallery's curator, Dr. Martin Collins, with whom we met several months before the exhibit's opening, "One World Connected" seeks to convey, within the constraints of a national aerospace museum, both the promises and vulnerabilities created by globalization, pointing to the unevenness and ambiguity of global interconnectedness.

FIGURE E.1
Promotional image of the "One World Connected" gallery. Reproduced by permission of the Smithsonian's National Air and Space Museum.

The "One World Connected" gallery continues to promote key ideas from the first decades of satellite communications, including the claim that media globalization can create an experience of live, global presence. The British, European, Japanese, American, and Soviet producers of 1967's "Our World" and "One Hour in the Life of the Motherland" broadcasts would recognize the ideas behind the photojournalistic panels and whole-Earth views on display in "One World Connected." The new exhibit's concerns about inclusion and exclusion, access and unevenness, and how to center a single country's technical contributions in the context of a global network are equally familiar, although Dr. Collins and his contracted team of British exhibit developers sought to go further in their inclusion of non-Western perspectives. Their exhibit, for example, includes a display featuring an American ATS-6 satellite, the model used to facilitate the US–India satellite education program, and seeks to convey the perspective of Indian officials and participants wary of renewed imperialism by new means.

The gallery's title, "One World Connected," certainly has a special resonance in the US, where the original dream of US telecommunications officials responsible for the creation of Intelsat was a "single global network," unified by US technology and outside the constraints of the International Telegraph

FIGURE E.2
Poster for "Satellite Communications. On Earth and in the Cosmos." Reproduced by permission of the Museum of Cosmonautics, Moscow.

Union's existing global media governance structures. Yet the National Air and Space Museum is not the only major world aerospace museum to attempt to incorporate applications satellite history—and new, for-profit satellite communications providers—into its galleries. Russia's national aerospace museum in Moscow, the Museum of Cosmonautics, likewise organized a major new exhibit on the history of communications satellites, entitled "Satellite Communications: On Earth and in the Cosmos" (figure E.2). Although the exhibit was mounted for only a year, it nonetheless represented a substantial investment of time and resources. "Satellite Communications: on Earth and in the Cosmos" occupied a large, central gallery, one through which all visitors pass, from November 2018 until November 2019.

The Museum of Cosmonautics exhibit differed from its American counterpart, but it too reflected a growing interest in both transnational relationships and the commercial history of space technology. The exhibit focused extensively on transnational interactions, including a panel dedicated to the history of the satellite Direct Communications Link (DCL) between

Moscow and Washington—also a key feature of the National Air and Space Museum's new gallery. In Moscow, detailed exhibition panels included both original objects and new technical drawings depicting moments and aspects of international cooperation in satellite communications that have received little scholarly attention in the English-language literature. These included, for example, plans for the satellite linkups that transmitted the July 1975 broadcast of the Apollo–Soyuz docking mission via Houston and Moscow to television viewers and radio listeners across the Soviet Union.

The Museum of Cosmonautics exhibit foregrounded the specific technical aspects of Soviet and international satellite network connections, as well as various Soviet firsts, rather than explicitly centering global or planetary visions. Nonetheless, common themes did emerge. Some featured archival objects that reflected the presence of planetary visions originating in Moscow and elsewhere in the Soviet Union, such as a small still from the first color television transmission of the view of Earth from space, made by a Molniya satellite in 1967. A map of the first Molniya satellite's trace and visibility zone likewise depicted the Molniya satellite's path right across the US and Canada in a global projection that resembled Intelsat coverage zone maps. Much as the "One World Connected" gallery highlights the role of private satellite firms in providing many contemporary applications satellite services, the Museum of Cosmonautics exhibit's main character was the Soviet and post-Soviet Russian satellite communications industry: the exhibit was sponsored by the Russian Satellite Communications Corporation, a state-controlled entity that inherited much of the Soviet communications satellite space array and ground infrastructure. The exhibit avoided larger questions about the unevenness or challenges of globalization and presented the history of satellite communications through the frame of corporate history, focusing on roles and contributions of satellite communications personnel, the ground infrastructures where they worked, and the formerly secret Soviet technical institutes that have now become private satellite equipment engineering and manufacturing firms. Television also plays a major role, reflecting the stronger association between communications satellites and broadcast media in Russia, where communications satellites brought the first countrywide television coverage to the Soviet Union in the late 1960s. The exhibit thus featured many specific examples of the global connections and planetary visions facilitated by Soviet satellites and the infrastructural integration they generated.

Unlike the Smithsonian "One World Connected" gallery, "Satellite Communications. On Earth and in the Cosmos" included no large globe or other central feature designed to emphasize global interconnection as the main story of satellite history. Instead, it featured a patchwork of moments of interconnection and the technical networks, objects, and people that made those moments possible. The exhibit's poster does present a view of the Earth from space, but the image centers on Eurasia and the Eastern Hemisphere, reminding us that no view from space can actually encompass the whole planet. If the promotional images of the new National Air and Space Museum gallery echoed the visual language of the 1967 "Our World" broadcast, with its large globe and emphasis on the view of Earth from space, the Museum of Cosmonautics exhibit poster draws on the imagery— also from 1967—of the first maps of the Orbita network of Earth stations. Like those maps, the poster concealed the dangerously transnational nature of satellite communications technology, including only Earth stations on Russian territory.

The Museum of Cosmonautics satellite communications exhibit closed more than two years before the second Russian invasion of Ukraine in February 2022, which has had far-reaching and devastating consequences for millions of Ukrainian civilians, including through the explicit targeting of transborder infrastructures including power grids, oil and gas pipelines, and internet data centers. The poster for the Museum of Cosmonautics "Satellite Communications: On Earth and in the Cosmos" exhibit reminds us that aspirations toward autarkic, exclusively national communications networks and hard borders, however factually inaccurate, were always a central part of how satellite communications were represented and sold to the global public from the 1960s onward. The exhibit's content, however, like the real work of satellite communications engineers and officials, offers a more complex picture, in which engineers, technical workers, and local communities around the world helped envision and construct an inherently planetary communications infrastructure.

Asymmetrical actors—those participants in the process of satellite network construction who commanded fewer resources, who were located on the margins of the earliest US satellite communications experiments, or who attempted to create alternative networks and nodes within an emergent planetary communications infrastructure—have been central to the narrative of this book. They have been involved in emerging satellite broadcast

practices, negotiated institutional belonging, constructed networks of satellite ground infrastructure to facilitate the distribution of satellite signals, and
been part of the integration of these networks on a planetary basis. Other
scholars have begun to document the substantial influence over space infrastructure, planetary conceptualizations, and human space activity in general
that has been wielded by officials and citizens in countries near the equator, whose land and participation have been essential to the construction of
infrastructure for human space activity. Our focus, based on our own back-
grounds in European and Soviet broadcasting history, has been on the role
of Eastern European socialist countries in shaping satellite communications
infrastructure, from the first transnational satellite television experiments
and global satellite broadcasts, such as 1967's "Our World," to the creation
of apparently separate and opposing Cold War satellite communications networks, Intelsat and Intersputnik, which in fact were interconnected from the
beginning and became more and more integrated over time. This integration took place thanks to shared Soviet and Western European interests in
profiting from commercial space communications and countering the US
influence, the relationships forged by the construction and promotion of
satellite Earth stations around the world, and the unintended consequences
of Cold War nuclear diplomacy, such as the decision to create a satellite DCL.
The history of global media infrastructure cannot be written without attending to the myriad, variously empowered and disempowered participants and
contributors—people, institutions, and planetary and highly local geographies and topographies—who helped shape the material and institutional
forms those infrastructures take.

Perhaps the ultimate demonstration that the US officials who built Intelsat were profoundly shaped by their interactions with global partners, clients,
and rivals was the way that Intelsat's leadership reacted when confronted, in
1984, with the greatest threat to the globalizing visions on which Intelsat was
founded—a threat that came from the US government itself. US officials, of
course, had earlier played a significant role in the gradual breakdown of US
control over Intelsat and in the failure to resist the formation of rival regional
networks in the 1970s, when they prioritized European participation in the
shuttle program over Intelsat, as we argued in chapter 5. The 1984 determination by US president Ronald Reagan to allow privately owned global communications satellites, however, was far more serious. Indeed, according to
Intelsat representatives themselves, it posed an existential threat. As Intelsat

officials saw it, if the satellite industry were transformed into a private market driven by competition, including in the most profitable transatlantic market, where Intelsat had previously enjoyed an officially guaranteed monopoly on US-international satellite traffic, Intelsat would no longer be profitable enough to function as a global system.

As it defended its special status within the US, Intelsat returned to its founding vision, as a single global network, recycling the rhetoric that it had adopted during the 1969–1971 permanent arrangement negotiations. Looking back to Intelsat's first decade, Abbott Washburn pointed out that Intelsat negotiations in the 1960s aimed at making regional systems compatible with Intelsat, and that *"transocean* systems were not contemplated" since it was evident to everyone that they were "in direct conflict with the single global system concept."[1] For Washburn and other Intelsat officials, US leadership in global communications was still central to the idea of a single global network, and opening the market to private firms would mean the destruction of the leading US position. To Washburn, Intelsat represented one of the "most successful and useful international initiatives of the U.S. in our century."[2] Joseph N. Pelton, executive assistant to the director general of Intelsat concluded his talk at the 1983 Satellite Communications Users Conference in St. Louis even more dramatically, petitioning his audience to "keep the wonderful baby the U.S. gave to the world so generously 18 years ago." This baby, he stressed, was facing an existential threat captured by the title of his talk: "If someone is dead, would you ask if they were significantly wounded?"[3]

Yet the risk of destroying a precious US global achievement was not the only rhetorical tactic employed by current and former US officials who had helped to build Intelsat in the 1960s and early 1970s. They also claimed that Intelsat's main task was to provide equitable, affordable satellite communications access to developing countries, drawing on rhetoric about fairness, equal access, and affordability that reflected less the original US goals for Intelsat than the mutually negotiated position forged during the Cold War with both the socialist bloc and nonaligned countries. When faced with the threat of private competition for transatlantic routes, in other words, Intelsat stressed its ostensible commitment to values of equality and fairness to the developing world that had originated as nonaligned and Soviet arguments about the need for an alternative to Intelsat. Pelton, in his 1983 speech on the issue, equated privatization with cutting off the developing world's access to global communications networks, stating that "scores of these countries are totally dependent upon INTELSAT as their only means of overseas communications."[4]

In his June 1984 testimony before the House Subcommittee on Telecommunications and Finance, Irving Goldstein, the president of COMSAT, went even further in linking Intelsat to the goals of communications access and equity.[5] Goldstein stressed the essential role of Intelsat's monopoly on the "lucrative, high-volume" routes across the North Atlantic in subsidizing access for developing countries, making possible "affordable and nondiscriminatory" rates for all.[6] Private communication satellite firms like Orion and other commercial operators, Goldstein stressed, would not be interested in serving "developing nations whose routes provide a much smaller profit potential."[7] Even as Goldstein attributed to Intelsat the same attributes of fairness and equal access that had originally been Intersputnik talking points, however, he continued to deploy the threat of Soviet expansion, via Intersputnik, into the developing world. Equitable and affordable pricing, made possible by Intelsat's monopoly over more lucrative transatlantic routes was needed precisely to prevent the Soviet Union from luring developing countries with the same low, equitable prices. "A weakened INTELSAT would place greater burdens on developing nations," Goldstein warned, "burdens that would make them vulnerable to entreaties from the Soviet Union and its rival international system, Intersputnik."[8] In a statement to the US Congress, Washburn likewise warned congressional representatives about Intersputnik's efforts to adopt a more competitive position, upgrading its system and showing a willingness to subsidize new members by providing free services and assistance with the construction of new Earth stations designed to receive signals from Intersputnik satellites.[9] Threatened by private firms like Orion, Intelsat thus defended itself as a bulwark against Intersputnik, but one that exemplified values of equity and fairness to the postcolonial world that were nearly indistinguishable from Soviet claims.

By the mid-1980s, Intelsat was challenged on multiple fronts, finding itself in a situation where the Reagan administration, as well as the Federal Communications Commission (FCC), had acted, as Intelsat saw it, to undermine its position as a global leader in satellite communications, abandoning an understanding of US satellite communications policy that had been shaped both by an imperializing US global telecommunications policy in the 1960s and by extensive interactions with the socialist world. Yet time appeared to have run out not only on this shared vision, but even on the use of Intersputnik as a bogeyman with which to mobilize congressional allies.

By the summer of 1984, Intersputnik was already making minor inroads into US domestic broadcasting, thanks to Ted Turner's Cable News Network (CNN). CNN's decision to broadcast the Friendship Games from Moscow in the summer of 1984 was followed by intense negotiations over the permanent use of the Intersputnik network to broadcast news and other content from the so-called Eastern bloc. In February 1985, CNN filed an application with the FCC for "approval of a permanent earth station in Atlanta to point at an INTERSPUTNIK satellite to bring the Soviet's Intervision TV news programming to the U.S. on a regular basis."[10] US officials' fears, articulated from the very beginnings of the plans for Intelsat in the early 1960s, about the relative ease with which Earth stations could be adapted to receive signals from another system's satellite, were now fully realized, and on the US's own territory. The culprit was not a developing country, whose poverty, American officials had condescendingly proposed, might make them fickle and vulnerable to Soviet influence. Instead, the breakdown of Intelsat's global network, and most crucially its monopoly on international satellite communications between the US and Europe, was led by American private-sector capitalists and media moguls.

Academic writing on Intelsat and Intersputnik from this era did not reflect either Intelsat's response to the end of its monopoly or Intersputnik's increasingly energetic attempts to expand its global reach not by displacing Intelsat, but by working alongside it or even using Intelsat's ground stations as infrastructure, as in the Managua Earth station. Instead, by the middle of the 1980s, American economists and political scientists who evaluated the state of Soviet space communications had returned to a binary, competitive framework typical of the Space Race twenty years earlier. Focused on comparing Intersputnik and Intelsat as discrete and rival networks, they found that Intersputnik was hopelessly technically behind, characterized by inefficient investment, inferior technology, and a weak competitive position relative to Intelsat's global network.[11] Both Intersputnik and the domestic Soviet communications satellite network, as the economist Robert Campbell argued, "seem to confirm our ideas about [Soviet] technological weaknesses and the flabbiness of innovative drive."[12]

From the perspective of the present, however, these bodily metaphors of Soviet flab and, implicitly, American fitness and vigor appear quite misplaced. The use, by American Sovietists of the 1980s, of a binary, competitive framework for assessing Soviet space activities concealed the reality of extensive Soviet engagement across Cold War boundaries and in global institutions, about which COMSAT officials warned during their response to the Orion challenge. The Soviet Union's expanded outreach and engagement with both global aerospace firms and developing world clients in the 1980s positioned post-Soviet Russian space agencies and infrastructures to compete more effectively in an increasingly diversified global satellite communications market after 1991.[13]

Space industry coverage of Intersputnik after 1991 was marked by a similar erasure of Intersputnik's engagement in space communications markets before the collapse of the Soviet Union, promoting a narrative that contrasted Soviet-era backwardness and isolation with a post-1991 capitalist Russia's desire for profit. An October 1992 article in *Space News* described Intersputnik as "once a sleepy Russian bureaucracy" that now had "taken its aging satellites and built a robust business," chiefly by offering "bargain basement prices."[14] Building a robust business by undercutting Intelsat's prices, of course, had been Intersputnik's objective from its initial formation in 1968. The article likewise presented Intersputnik's status as an international membership organization that employed many non-Russian citizens in its administrative structure as something new and surprising, and also described Intersputnik member-countries as "former Soviet strongholds," flattening the complexity of Intersputnik's relationships with nonaligned countries, as well as its infrastructural integration with Intelsat networks in those countries and, indeed, on Soviet territory itself. For Western audiences, Intersputnik's original objectives—commercializing Soviet space technology for financial gain and building an international satellite communications network that could both compete and cooperate with Intelsat—finally made sense in the context of Russia's capitalist transition. Intersputnik's quite consistent institutional structure and mission were thus rediscovered by Western journalists after 1991 and reinvented as the product of the collapse of communism.

Recovering the story of Soviet and socialist world participation in and influence on the creation of satellite communications institutions and infrastructures in the era of decolonization and Cold War might seem to have little significance. And yet many of the anxieties about the threatening potential of global media networks and infrastructures that were expressed by participants in the development of satellite communications have remained quite salient up to the present day. Recent years have witnessed a new controversy over yet another global communications network, the development of 5G telecommunications infrastructures that provide internet broadband access on a global scale. Rather than the US and the Soviet Union (or Russia, after 1991), however, the two main antagonists in this controversy have been the US and China, with US concern focusing on the relationship between Beijing and Huawei, the Chinese firm that produces much of the equipment used to build and administer new 5G network infrastructures. Concerned about the risk of Chinese government espionage and control over critical infrastructure in the US and elsewhere, the US government in 2020 banned the use of Huawei equipment, pressing other countries to follow suit.[15]

The US government's position in the case of 5G networks is thus diametrically opposed to the position that it took in the first decades of Intelsat's formation and expansion. Instead of supporting the vision of a single, monopolistic global operator of media infrastructures, the US, in the case of 5G networks, has occupied a position much like that of its former rivals, France or even the Soviet Union itself. It proposed that 5G technology be organized into multiple fragmented and highly regionalized global networks rather than allowing a single party to predominate. The stakes of this conflict are largely the same, however, centering on questions of access and global connectivity, as well as fears over the control and surveillance of critical infrastructures and the distribution of the profits from manufacturing and construction of the material network itself. The shift in the US position reflects the more multicentered nature of 5G networks, but also the real decline in US technological and geopolitical dominance. The US loss of market dominance in telecommunications technology was already evident during the expansion of satellite communications infrastructures in the mid-1970s, as non-US firms built Intelsat Earth stations around the world, and continues today; the largest US telecommunications manufacturing company, Cisco, ranks fifth among global telecommunications firms.[16]

Debates about media globalization and internet governance have likewise raised renewed concerns about governance and control over media infrastructures. Early debates over internet governance offered polarized accounts of a global internet either best governed by freedoms ostensibly inherent to the technology itself, or by existing models of government control sought by advocates of a "bordered internet."[17] While the bulk of internet traffic historically has used terrestrial and undersea cables, the recent introduction of satellite internet services employing so-called microsatellites in low-Earth orbit has reasserted the connections between the internet and satellite communications. Satellite internet providers such as Viasat, HughesNet, and Starlink address the parts of the globe that have remained excluded from cable-based high-speed internet, primarily serving rural and remote areas and offering broadband connectivity to users where cable networks are sparse.[18] Starlink, a unit within Elon Musk's SpaceX company, has placed satellites in low-Earth orbit to allow lower latency (i.e., the time needed for the signal to travel between Earth and satellites). This offers a faster internet connection, but at the cost of not being able to place satellites in geosynchronous orbits and providing a significantly smaller satellite footprint. To increase the coverage, providers such as Starlink have launched a very large number of satellites, with over 4,000 satellites since 2018 by Starlink alone.[19] This has raised concern about the overpopulation of orbital routes and potential of causing outer space collisions, such as when the Tianhe, a Chinese space station module, had to maneuver to avoid being hit by Starlink satellites in 2021.[20] The large number of satellites have also distorted telescope images and impeded astronomical research.[21] Most news coverage of satellite internet has thus focused on the space segment of the network, not least in light of the serious problems of space debris and crowding of satellite orbits.

These developments, however, have also put communications satellite ground infrastructure at the center of attention.[22] After Russia launched its war on Ukraine in late February 2022, it soon became evident that shelling and cyberattacks posed a great threat to Ukraine's internet infrastructure, prompting SpaceX to provide the country with Starlink terminals, although Musk soon wavered in his commitment to supporting Ukraine and complained petulantly about the cost of that assistance.[23] The terminals resemble rooftop antennas used for direct broadcast television, and almost one month into the war, it was reported that over 5,000 terminals had been shipped into the country.[24]

In addition to mimicking the early promises of internet, and particularly Arpanet, of being a distributed communications network, and thereby less prone to outage and disruption, the language used on Starlink's own web page echoes that of earlier satellite networks, such as Intelsat, in that it promises "truly global coverage," reaching "far more people and places," in addition to being "much faster than in fiber-optic cable."[25] Just like the first communications satellite networks, Intelsat and Intersputnik, Starlink promised access to a global modernity, as well as contributing to a sense of global unification as they announce their aim to "close the rural broadband gap." Yet, as with earlier, geosynchronous satellite communications systems, establishing the ground infrastructure needed to receive and distribute signals from space has turned out to be both essential and far more challenging than anticipated, facing both technical and political challenges on the Earth's surface.

Once again, moreover, promises of global coverage and unification have been subject to substantial challenges and criticism. In the 2020s, it is no longer Intelsat and Intersputnik who are the main actors in satellite communication, and rather than the promises and dangers of global television broadcasting, it is instead global internet via satellites that is at the center of our attention. Yet the core issues and debates remain quite similar. SpaceX and Starlink are far from the only companies looking to establish a satellite constellation in low-Earth orbit; they face competition from companies in the UK, US, and China. In a research report outlining the consequences of satellite constellations such as Starlink, Daniel Voelsen of the German Institute for International and Security Affairs, or Stiftung Wissenschaft und Politik (SWP), described a situation strongly reminiscent of 1960s European evaluations of the threat of US dominance of global satellite communications. Voelsen suggested that future internet governance initiatives must prepare for two possible scenarios.[26] The first would be the state of "global oligopolies," where the economic and political power is concentrated to three "satellite mega constellations," two controlled by the US and UK, whereas the third would be part of the Belt and Road Initiative, China's global infrastructure project.[27] In such a scenario, Voelsen argues, the "internet further fragments," and the countries behind the constellations will have "fine-grained control over exactly how data . . . is exchanged," leaving European countries "powerless to shape the use of digital infrastructures."[28] Voelsen's second scenario entails "regulated competition" under the auspices of, for example, the World Trade Organization, which would force companies operating satellite megaconstellations to cooperate with local companies and service providers. This would, Voelsen argued, allow European countries to foster a "close technological partnership with Japan," as well as cooperation between the European Union and African

Union in securing affordable internet for a large number of people in developing countries.

This scenario would produce an alternative to US and Chinese systems, and, Voelsen admitted, would see the constellations partly become an "instrument of vested geopolitical interests," while at the same time preserving "the common global foundation of the Internet."[29] Voelsen's proposal, with its concerns over European influence within media governance institutions and its mobilization of concern for access and equity for citizens of the Global South, echoes the rhetoric of European negotiators in the late 1960s and 1970s, as they sought to ensure Europe's role in controlling and profiting from global media networks. The uncertainty characterizing the future of media infrastructures in the 2020s thus bears a striking resemblance to the history of the first two decades of satellite communications. While the actors may be different, the effort to resolve the "tension between local and global," to borrow Star and Ruhleder's phrase, remains central.[30]

The history of satellite communications in the era before direct broadcast satellites thus offers important insights into the recurring patterns of global media infrastructural development. Yet we hope to have shown that this is not a story of unidirectional technology transfer and global expansion, led by a triumphant postwar US, and neither is it one of just a few Western actors creating and gradually expanding transnational broadcast experiments into an experimental satellite network and then a global media infrastructure. Rather, the process of becoming infrastructure can only be captured by a multifaceted history of the coming-together of technologies, broadcast practices, transnational institutions, and terrestrial networks. The communications satellite networks of the 1960s and 1970s were formed by mutual interaction, mimicry, and often shared understandings of how satellite communications technology should be developed and institutionalized. These interdependencies, however, are easily forgotten in favor of stories of national achievement that conceal the often anxiety-provoking and boundary-transgressing processes of global media infrastructure construction. Looking at satellite communications infrastructures not from space, but from the thousands of places on Earth where satellite signals were sent, received, and passed along via terrestrial networks, reminds us of the horizontality of space media and suggests that we pay close attention to histories of media infrastructures since they still inform and reflect our present beliefs, desires, and fears about global media infrastructures.

TERMS AND ABBREVIATIONS

CNES
Centre Nationale des Etudes Spatiales (National Center for Space Studies). CNES, is the French government's national space agency, created in 1961.

COMSAT
Communications Satellite Corporation. Created by an act of the US Congress as a government-regulated private corporation authorized to develop commercial communications satellite technology. COMSAT also served as the US representative to Intelsat and Inmarsat.

Earth station or ground station
A terrestrial radio station featuring a large parabolic antenna capable of sending and receiving radio signals to and from spacecraft, such as a communications satellite in orbit.

EBU
European Broadcasting Union. An association of European public service broadcasters created in 1950 and headquartered in Geneva. The EBU administers the Eurovision network, which fosters the exchange of television and radio broadcasts. Between 1950 and 1993, most members were Western European broadcasters; former International Organization for Radio and Television (OIRT) members joined the EBU in 1993.

ESA
European Space Agency. An intergovernmental membership organization, headquartered in Paris and dedicated to space exploration. Created in 1975 on the basis of a merger between the European Launch Development Organization (ELDO) and the European Space Research Organization (ESRO).

IBU

International Broadcasting Union. An association of European broadcasters created in 1925 and headquartered in Geneva. Predecessor to both the International Organization for Radio and Television (OIRT) and the European Broadcasting Union (EBU). Plagued by conflicts after World War II, with OIRT broadcasters leaving the association in 1946. Dissolved in May 1950, after the foundation of EBU in February the same year.

Intelsat

International Telecommunications Satellite Organization, headquartered in Washington, D.C. From its creation in 1964 until its privatization in 2001, an intergovernmental organization that owned and operated communications satellites and provided international commercial satellite communications services, including telephony, data transmission, and television broadcasting.

Intersputnik

An intergovernmental satellite communications organization, headquartered in Moscow, that provided international commercial satellite communications services including telephony, data transmission, and television broadcasting. Plans for Intersputnik were announced in 1968, and the organization was formally initiated in 1971.

ITU

International Telecommunications Union. A United Nations (UN) agency, headquartered in Geneva, tasked with regulating and coordinating global telecommunications and information technologies, including shared use of the radio spectrum and the allocation of orbital positions for satellites.

Molniya

A series of communications satellites produced and launched by the Soviet Union beginning in 1965. Employing a highly elliptical orbit, Molniya satellites offered better coverage of the Soviet Union's far northern territories.

OIR/OIRT

Organization Internationale de Radiodiffusion et de Télévision (International Radio and Television Organization). Originally OIR, with the "T" (for television) added in 1960. An association of chiefly East European public service broadcasters created in 1946 and headquartered in Prague. The OIRT administered the Intervision network, which fostered the exchange of television and radio broadcasts among member-countries. In 1993, pursuant to the collapse of the Soviet Union, the OIRT merged with the EBU.

Orbita

The Soviet domestic satellite communications system, which was initiated in 1967 and brought national television broadcasting service to the majority of Soviet territory.

Orbita Earth stations across the Soviet Union sent and received radio transmissions from Molniya-series satellites in elliptical orbit.

Tracking station
An Earth station that communicates with or receives data from manned or unmanned space missions or satellites that are not in geostationary orbit.

NOTES

INTRODUCTION

1. Hannah Arendt, *The Human Condition*, 2nd ed. (Chicago: University of Chicago Press, 1958), 1.

2. Arendt, *The Human Condition*, 1. For more on Arendt's (and also Martin Heidegger's) idea of Earth alienation see Kelly Oliver, *Earth and World: Philosophy after the Apollo Missions* (New York: Columbia University Press, 2015); David Macauley, "Out of Place and Outer Space: Hannah Arendt on Earth Alienation: An Historical and Critical Perspective," *Capitalism Nature Socialism* 3: 4 (1992), 19–45; Marshall McLuhan likewise later described the launch of Sputnik as the end of nature, and the birth of ecology, based on the observation that for the first time, "the natural world was completely enclosed in a man-made container." For McLuhan, Planet Earth turned into Spaceship Earth, a global theater where all humans were actors rather than spectators, crew rather than passengers. Marshall McLuhan, "At the Moment of Sputnik the Planet Became a Global Theater in Which There Are No Spectators but Only Actors," *Journal of Communication* 24: 1 (1974), 49. McLuhan's account draws on Buckminster Fuller's *Operation Manual for Spaceship Earth* (Carbondale: Southern Illinois University Press, 1969). But it should also be noted that Arendt, together with thinkers like Martin Heidegger and Hans Blumenberg, also considered Sputnik and later efforts in space exploration as a transformation of earth into a man-made planet. Benjamin Lazier, "Earthrise; or, the Globalization of the World Picture," *American Historical Review* 26 (2011), 604.

3. Nicholas Barnett, "'RUSSIA WINS SPACE RACE': The British Press and the Sputnik Moment, 1957," *Media History* 19: 2 (2013), 182–195. For an account of the involvement of radio amateurs in tracking Sputnik, see W. Patrick McCray, *Keep Watching the Skies! The Story of Operation Moonwatch and the Dawn of the Space Age* (Princeton, NJ: Princeton University Press, 2008); Veronica della Dora, "From the Radio Shack

to the Cosmos: Listening to Sputnik during the International Geophysical Year (1957–1958)," *Isis* 114: 1 (2023), 123–149.

4. Arendt, *The Human Condition*, 264; Hannah Arendt, "The Conquest of Space and the Stature of Man," *The New Atlantis. A Journal of Technology and Society* 18 (Fall 2007), 53.

5. For more on the "Blue Marble," "Earthrise," and representations of Earth, see Denis Cosgrove, *Apollo's Eye: A Cartographic Genealogy of the Earth in the Western Imagination* (Baltimore: Johns Hopkins University Press, 2001); Benjamin Lazier, "Earthrise; or, The Globalization of the World Picture," *American Historical Review* 116: 2 (2011), 602–630; Lisa Messeri, "The Moon's Earth," in Jeffrey S. Nesbit and Guy Trangos (eds.), *New Geographies 11, Extraterrestrial* (2020), 77–83; Robert Poole, *Earthrise: How Man First Saw the Earth* (New Haven, CT: Yale University Press, 2008); Howard Caygill, "Heidegger and the Automatic Earth Image," *Philosophy Today* 65: 2 (2021), 325–338; Fred Spier, "On the Social Impact of the Apollo 8 Earthrise Photo, or the Lack of It?" *Journal of Big History* 3: 3 (2019), 157–189; Chris Russill, "Guest Editorial: Earth-Observing Media," *Canadian Journal of Communication* 38: 3 (2013), 277–284; Mette Bryld and Nina Lykke, *Cosmodolphins: Feminist Cultural Studies of Technology, Animals and the Sacred* (London: Zed Books, 2000).

6. Andrew Jenks, "Securitization and Secrecy in the Late Cold War: The View from Space," *Kritika: Explorations in Russian and Eurasian History* 21: 3 (2020), 667. Other authors have pointed to the differences in the coverage and reception of the Apollo 8 mission, as well as the "Earthrise" photograph, noticing a "considerable cultural divide" both within and between Europe and the US, while still acknowledging its impact over time, he notes that it affected people in the US to a larger degree than in Europe. Spier, "On the Social Impact of the Apollo 8 Earthrise Photo," 183; see also Neil M. Maher, *Apollo in the Age of Aquarius* (Cambridge, MA: Harvard University Press, 2017); Sheila Jasanoff, "Heaven and Earth: The Politics of Environmental Images," in Marybeth Long and Sheila Jasanoff (eds.), *Earthly Politics: Local and Global in Environmental Governance* (Cambridge, MA: MIT Press, 2004); Sheila Jasanoff, "Image and Imagination: The Formation of Global Environmental Consciousness," in Clark A. Miller and Paul N. Edwards (eds.), *Changing the Atmosphere: Expert Knowledge and Environmental Governance* (Cambridge, MA: MIT Press, 2001), 309–337.

7. Lisa Messeri, *Placing Outer Space: An Earthly Ethnography of Other Worlds* (Durham, NC: Duke University Press, 2016). Most accounts of planetary imagination recognize the Cold War space programs and budding environmentalism as a starting point of planetary imagination. However, for a discussion of planetary thinking in early-twentieth-century Russia, see Daniela Russ, "'Socialism Is Not Just Built for a Hundred Years': Renewable Energy and Planetary Thought in the Early Soviet Union (1917–1945)," *Contemporary European History* 31: 4 (2022), 491–508.

8. Messeri, *Placing Outer Space*, 191. On seeing extraterritorial spaces such as Mars, see Janet Vertesi, *Seeing like a Rover: How Robots, Teams, and Images Craft Knowledge of Mars* (Chicago: University of Chicago Press, 2015); Peter Galison and Elisabeth Kessler, "To

See the Unseeable: Peter Galison in Conversation with Elizabeth Kessler," *Aperture* 237 (2019), 72–77.

9. Oliver, *Earth and World*, 20. However, drawing upon the mid-nineteenth-century practice of so-called photosculpture, Alexander R. Galloway argues that "there is an alternate history of photography *in which point of view has no meaning*, at least not a single point of view." Alexander R. Galloway, *Uncomputable: Play and Politics in the Long Digital Age* (London: Verso, 2021), 21 (emphasis in original). On satellite imaging, technologies of global imagery, and the relation between the true and the virtual, see Laura Kurgan, *Close up at a Distance: Mapping, Technology, and Politics* (New York: Zone Books, 2013).

10. Lisa Parks, *Rethinking Media Coverage: Vertical Mediation and the War on Terror* (London: Routledge, 2018), 9. See also Stephen Graham, *Vertical: The City from Satellites to Bunkers* (London: Verso Books, 2016).

11. Oliver, *Earth and World*; Messeri, *Placing Outer Space*.

12. Lisa Parks, "Satellites, Oil, and Footprints: Eutelsat, Kazat, and Post-Communist Territories in Central Asia," in Lisa Parks and James Schwoch (eds.), *Down to Earth: Satellite Technologies, Industries, and Cultures* (New Brunswick, NJ: Rutgers University Press, 2012), 122–137.

13. In *Down to Earth*, Lisa Parks and James Schwoch reverse the vertical launch and disappearance of satellites and foregrounds the "material and territorial relations of satellite technologies, industries, and cultures." A recent collection of essays makes a similar move, merging social studies of outer space with perspectives on infrastructure. Lisa Parks and James Schwoch, *Down to Earth: Satellite Technologies, Industries, and Cultures* (New Brunswick, NJ: Rutgers University Press, 2012), 1; Christine Bichsel, "Introduction: Infrastructure On/Off Earth," *Roadsides*, Collection no. 3 (2020), 1.

14. Judith T. Kildow, *INTELSAT: Policy-Maker's Dilemma* (Lexington, MA: Lexington Books, 1973); Olof Hultén, "The Intelsat System: Some Notes on Television Utilization of Satellite Technology," *International Communication Gazette* 19: 1 (1973), 29–37; Drew McDanie and Lewis A. Day, "INTELSAT and Communist Nations' Policy on Communications Satellites," *Journal of Broadcasting* 18: 3 (1974), 311–322; Marcellus S. Snow, "INTELSAT: An International Example," *Journal of Communication* 30: 2 (1980), 147–156; Hugh R. Slotten, "Satellite Communications, Globalization and the Cold War," *Technology and Culture* 43: 2 (2002), 315–350; Jill Hills, *Telecommunications and Empire* (Urbana: University of Illinois Press, 2007); Ingrid Volkmer, "Satellite Cultures in Europe: Between National Spheres and a Globalized Space," *Global Media and Communication* 4: 3 (2008), 231–244; Sarah Nelson, "Networking Empire: International Organizations, American Power, and the Struggle over Global Communications in the 20th Century" (PhD dissertation, Vanderbilt University, Nashville, 2021); Hugh R. Slotten, *Beyond Sputnik and the Space Race: The Origins of Global Satellite Communications* (Baltimore: Johns Hopkins University Press, 2022).

15. For more on Cold War media as characterized by extensive interaction and mutual influence across the Iron Curtain, see Alice Lovejoy and Mari Pajala, eds., *Remapping the Cold War Media: Institutions, Infrastructures, Translations* (Bloomington: Indiana University Press, 2022). For an excellent example of recent work highlighting interaction and contestation between the Global North and Global South in the early years of satellite communications technology, see Sarah Nelson, "A Dream Deferred: UNESCO, American Expertise, and the Eclipse of Radical News Development in the Early Satellite Age," *Radical History Review*, 141 (2021), 30–59.

16. Jennifer Gabrys, *Program Earth: Environmental Sensing Technology and the Making of a Computational Planet* (Minneapolis: University of Minnesota Press, 2016).

17. Asif Siddiqi, "Another Space: Global Science and the Cosmic Detritus of the Cold War," in Pedro Ignacio Alonso (ed.), *Space Race Archaeologies: Photographs, Biographies, and Design* (Berlin: DOM Publishers, 2016), 21–38; Lisa Parks, "Global Networking and the Contrapuntal Node: The Project Mercury Earth Station in Zanzibar, 1959–64," *Zeitschrift für Medien- und Kulturforschung* 11: 1 (2020), 40–57; Pedro Ignacio Alonso and Hugo Palmarola, "NASA in Chile: Technology and Branding of a Satellite-Tracking Station," *Design Issues* 33: 2 (2017), 31–42; Peter Redfield, *Space in the Tropics: From Convicts to Rockets in French Guiana* (Berkeley: University of California Press, 2000).

18. On these Franco-Soviet joint experiments, see Isabelle Gourne, "De Passer les Tensions Est-Ouest pour la Conquête de l'espace: La Coopération Franco-Soviétique au Temps de la Guerre Froide," *Cahiers SIRICE* 2: 16 (2016), 49–67; Jenks, "Securitization and Secrecy."

19. Lisa Parks and Nicole Starosielski, *Signal Traffic: Critical Studies of Media Infrastructures* (Urbana: University of Illinois Press, 2015), 4–5.

20. For similar approaches, see Lisa Parks, *Cultures in Orbit: Satellites and the Televisual* (Durham, NC: Duke University Press, 2005); Nicole Starosielski, *The Undersea Network* (Durham, NC: Duke University Press, 2015); Brian Larkin, *Signal and Noise: Media, Infrastructure, and Urban Culture in Nigeria* (Durham, NC: Duke University Press, 2008); Tung-Hui Hu, *A Prehistory of the Cloud* (Cambridge, MA: MIT Press, 2015).

21. Susan Leigh Star and Karen Ruhleder, "Steps toward an Ecology of Infrastructure: Design and Access for Large Information Spaces," *Information Systems Research* 7: 1 (1996), 113. See also Susan Leigh Star, "The Ethnography of Infrastructure," *American Behavioral Scientist* 43: 3 (1999), 377–391; Geoffrey C. Bowker and Susan Leigh Star, *Sorting Things Out: Classification and Its Consequences* (Cambridge, MA: MIT Press, 2000).

22. Brian Larkin, "The Politics and Poetics of Infrastructure," *Annual Review of Anthropology* 42: 1 (2013), 336.

23. For a discussion on the politics of making infrastructures invisible, see Lisa Parks, "Around the Antenna Tree: The Politics of Infrastructural Visibility," *Flow* 9: 8 (2009), 1–9.

24. Star and Ruhleder, "Steps toward an Ecology of Infrastructure," 114.

25. Nikhil Anand, Akhil Gupta, and Hannah Appel (eds.), *The Promise of Infrastructure* (Durham, NC: Duke University Press, 2018).

26. Joseph Vogl, "Becoming-Media: Galileo's Telescope," *Grey Room* (2008), 16; Bowker and Star, *Sorting Things Out*, 34.

27. Vogl, "Becoming-Media," 16.

28. James T. Andrews and Asif A. Siddiqi, *Into the Cosmos: Space Exploration and Soviet Culture* (Pittsburgh: University of Pittsburgh Press, 2011), 6.

29. See, for instance, Herbert I. Schiller, *Mass Communications and American Empire* (Boston: Beacon Press, 1969).

30. Kildow, *INTELSAT*; Snow, "INTELSAT"; John Downing, "The Intersputnik System and Soviet Television," *Soviet Studies* 37: 4 (1985), 465–483; John Downing, "International Communications and the Second World: Developments in Communication Strategies," *European Journal of Communication* 4: 1 (1989), 99–119.

31. F. B. Schick, "Space Law and Communication Satellites," *Western Political Quarterly* 16: 1 (1963), 14–33; Andrew Gallagher Haley, *Space Law and Government* (New York: Appleton-Century-Crofts, 1963); Edward G. Lee, "UNESCO Meeting on Space Communications: Legal Issues," *University of Toronto Law Journal* 20: 3 (1970), 375–379. For an updated discussion of international space law, the UN Committee on the Peaceful Uses of Outer Space, and the Outer Space Treaty, see Elina Morozova and Yaroslav Vasyanin, "International Space Law and Satellite Telecommunications," *Oxford Research Encyclopedia: Planetary Science*, 2019.

32. See David Whalen, *The Origins of Satellite Communications, 1945–1965* (Washington, DC: Smithsonian Institution Press, 2002); and Slotten, *Beyond Sputnik*.

33. Nelson, "Dream Deferred," 45–46.

34. Kaarle Nordenstreng and Tapio Varis, "Television Traffic: A One-Way Street? A Survey and Analysis of the International Flow of Television Programme Material," *UNESCO Reports and Papers on Mass Communication*, 1974; see also Tapio Varis, "The International Flow of Television Programs," *Journal of Communication* 34: 1 (1984), 143–152; and Tapio Varis, "Trends in International Television Flow," *International Political Science Review/Revue Internationale de Science Politique* 7: 3 (1986), 235–249. Concerns over imbalance in communication flows intensified with the dispute over so-called direct-broadcast satellites, allowing broadcasters to reach the homes of television viewers. Kathryn M. Queeney, *Direct Broadcast Satellites and the United Nations* (Alphen aan den Rijn: Sijthoff and Noordhoff, 1978).

35. Cf. Jean K. Chalaby, ed., *Transnational Television Worldwide: Towards a New Media Order* (London: Taurus, 2005).

36. Thomas P. Hughes, *Networks of Power: Electrification in Western Society, 1880–1930* (Baltimore: Johns Hopkins University Press, 1983); Wiebe E. Bijker, Trevor J. Pinch,

and Thomas P, Hughes, *The Social Construction of Technological Systems* (Cambridge, MA: MIT Press, 1987); Paul N. Edwards, *A Vast Machine: Computer Models, Climate Data, and the Politics of Global Warming* (Cambridge, MA: MIT Press, 2010); Larkin, *Signal and Noise*.

37. Kildow, *INTELSAT*, 7.

38. Diana Lemberg, *Barriers Down: How American Power and Free-Flow Policies Shaped Global Media* (New York: Columbia University Press, 2019).

39. The closeness of the fantasy of communication as a "project of reconciling self and other" and of persuasion and management of opinion is pointed out by John Durham Peters, saying that communication as "reduplication of the self . . . in the other . . . is in essence . . . [an attack on] the distinctiveness of human beings." John Durham Peters, *Speaking into the Air: A History of the Idea of Communication* (Chicago: University of Chicago Press, 1999), 9, 21.

40. The multiple and often-conflicting promises of infrastructure as modernity has often been addressed in relation to colonialism, modernization, and infrastructures, see Larkin, *Signal and Noise*; Larkin, "The Politics and Poetics of Infrastructure"; Anand et al., *The Promise of Infrastructure*.

41. See, for example, the essays in Alexander C. T. Geppert (ed.), *Limiting Outer Space: Astroculture after Apollo* (London: Palgrave Macmillan, 2018); Gemma Cirac-Claveras, "Re-Imagining the Space Age: Early Satellite Development from Earthly Fieldwork Practice," *Science as Culture* 31: 2 (2022), 163–186.

42. Alexander C. T. Geppert, "The Post-Apollo Paradox: Envisioning Limits during the Planetized 1970s," in Alexander C. T. Geppert (ed.), *Limiting Outer Space: Astroculture after Apollo* (London: Palgrave Macmillan, 2018), 3–26. To Arendt and Heidegger, space activities highlighted not only the entangled relation between space and globalization, but also the dangers of the latter. See Oliver, *Earth and World*, 26.

43. Paul N. Edwards, "Meteorology as Infrastructural Globalism," *Osiris* 21: 1 (2006): 230; Edwards, *A Vast Machine*. Edwards developed his concept of infrastructural globalism based on Martin Hewson's work on informational globalism. See Martin Hewson, "Did Global Governance Create Informational Globalism?" in Martin Hewson and Timothy J. Sinclair (eds.), *Approaches to Global Governance Theory* (Albany, NY: SUNY, 1999), 97–113. Mariel Borowitz, in a similar fashion, observes that the earliest uses of the term "open data" was in relation to Earth-observing satellites and remote sensing, and meteorological data was shared between US and the Soviet Union in the early 1960s. Mariel Borowitz, *Open Space: The Global Effort for Open Access to Environmental Satellite Data* (Cambridge, MA: MIT Press, 2017), 5, 239.

44. Parks, "Global Networking and the Contrapuntal Node"; Redfield, *Space in the Tropics*; Siddiqi, "Another Space"; Asif Siddiqi, "Shaping the World: Soviet-African Technologies from the Sahel to the Cosmos," *Comparative Studies of South Asia, Africa and the Middle East* 41: 1 (2021), 41–55.

45. Andrew Jenks, "Transnational Utopias, Space Exploration and the Association of Space Explorers, 1972–85," in Alexander C. T. Geppert (ed.), *Limiting Outer Space: Astroculture after Apollo* (London: Palgrave McMillan, 2018), 209–210.

46. Pedro Ignacio Alonso, "Introduction: Towards an Archaeology of Things Moving," in Pedro Ignacio Alonso (ed.), *Space Race Archaeologies: Photographs, Biographies, and Design* (Berlin: DOM Publishers, 2016), 17. The broader geography of space historical research mirrors a growing interest in other extraterritorial spaces, such as the ocean as a site for offsetting the terrestrial bias that have informed and defined our conceptualization of media. "Thinking through seawater," says Melody Jue in her study of ocean media, allows for challenging notions that have been taken for granted despite being milieu-specific to life out of water. Melody Jue, *Wild Blue Media: Thinking through Seawater* (Durham, NC: Duke University Press, 2020). See also Melody Jue and Rafico Ruiz (eds.), *Saturation: An Elemental Politics* (Durham, NC: Duke University, 2021); John Shiga, "Sonar: Empire, Media, and the Politics of Underwater Sound," *Canadian Journal of Communication* 38: 3 (2013), 357–377.

47. Gabrys, *Program Earth*; cf. Edward Jones-Imhotep, *The Unreliable Nation: Hostile Nature and Technological Failure in the Cold War* (Cambridge, MA: MIT Press, 2017); Sebastian Vehlken, Christina Vagt, and Wolf Kittler, "Introduction: Modeling the Pacific Ocean," *Media+Environment* 3: 2 (2021); Adam Wickberg and Johan Gärdebo, (eds.), *Environing Media* (London: Routledge, 2022).

48. Siddiqi, "Shaping the World," 50.

49. Here, we use the planetary rather than the global. The latter term is used by the actors involved in developing communications satellite networks, often describing them in singular form as "a global system of satellite communications," highlighting how the notion of the global make claims of universality and "expansive flattening"; cf. Messeri, *Placing Outer Space*, 10. When discussing a wider conceptualization of not only the "global communication networks," but a reshaping of thoughts, ideas and actions beyond the networks themselves, we instead use the term planetary.

50. Jenks, "Securitization and Secrecy"; Gourne, "De Passer les Tensions Est-Ouest."

51. Per Högselius, *Red Gas: Russia and the Origins of European Energy Dependence* (New York: Palgrave Macmillan, 2013); the concept of "hidden integration" was introduced by scholars of history of technology within the Tensions of Europe network. See Thomas J. Misa and Johan Schot, "Inventing Europe: Technology and the Hidden Integration of Europe," *History and Technology* 21: 1 (2005), 1–19, as well as the other articles in this special issue. The Tensions of Europe network has resulted in a large number of articles and books, not least on communication. Hidden integration, however, has primarily been developed in relation to other infrastructures and network such as waterways, railways, electricity grids, and natural gas pipelines. See Per Högselius and Yao Dazhi. "The Hidden Integration of Eurasia: East-West Relations in the History of Technology," *Acta Baltica Historiae et Philosophiae Scientiarum* 5: 2 (2017), 71–99; Per Högselius, "The Hidden Integration of Central Asia: The Making

of a Region through Technical Infrastructures," *Central Asian Survey* 41: 2 (2021), 223–243 .

52. See, for example, Simo Mikkonen and Pia Koivunen, *Beyond the Divide: Entangled Histories of Cold War Europe* (New York: Berghahns Books, 2015); Peter Romijn, Giles Scott-Smith, and Joes Segal (eds.), *Divided Dreamworlds? The Cultural Cold War in East and West* (Amsterdam: Amsterdam University Press, 2012).

53. James Mark, Artemy Kalinovsky, and Steffi Marung (eds.), *Alternative Globalizations: Eastern Europe and the Postcolonial World* (Bloomington: Indiana University Press, 2020), 2.

54. See, for example, Sari Autio-Sarasmo, "Stagnation or Not? The Brezhnev Leadership and the East-West Interaction," in Dina Fainberg (ed.), *Reconsidering Stagnation in the Brezhnev Era: Ideology and Exchange* (Lanham, MD: Lexington Books, 2016); Sari Autio-Sarasmo, "Technological Modernisation in the Soviet Union and Post-Soviet Russia: Practices and Continuities," *Europe-Asia Studies* 68: 1 (2016), 79–96. See also Högselius, *Red Gas*.

55. Benjamin Peters, *How Not to Network a Nation: The Uneasy History of the Soviet Internet* (Cambridge, MA: MIT Press, 2016).

56. Gabrielle Hecht, "Introduction," in Gabrielle Hecht (ed.), *Entangled Geographies: Empire and Technopolitics in the Global Cold War* (Cambridge, MA: MIT Press, 2011), 2.

57. Special thanks to Professor Bohdan Shumylovych for his assistance and for sharing his own work on the Lviv Earth station. See Bohdan Shumylovych (text) and Olha Povoroznyk (video), "Future from the Past: Imaginations on the Margins," https://ars.electronica.art/keplersgardens/de/imaginations/, September 12, 2020. (accessed March 27, 2023).

58. For example, when we worked with folders covering the BBC's relation with broadcasters in socialist countries, including the Soviet Union, East Germany, and Poland, a large number of them had never been accessed by researchers before. This is despite the fact that the BBC Written Archives Centre is a major archive for television historians not only in the UK, but also more widely.

59. Parks, *Cultures in Orbit*.

60. Marybeth Long Martello and Sheila Jasanoff characterize these new forms of governance as featuring "the increasing interaction between scientific and political authority highlighting fault lines in each; the salient role of non-state actors in both knowledge making and politics; the emergence of new political forms in response to novel conjunctions of actors, claims, ideas, and events that cut across national boundaries." "Introduction: Globalization and Environmental Governance," in Marybeth Long Martello and Sheila Jasanoff (eds.), *Earthly Politics: Local and Global in Environmental Governance* (Cambridge, MA: MIT Press, 2004), 1–29.

CHAPTER 1

1. In the recording from the Swedish television broadcast, three layers of commentary can be heard simultaneously, the original Russian, hardly audible in the background, and the British and Swedish commentary directed at the viewers. "Kosmonauten Gagarins ankomst till Moskva" [Cosmonaut Gagarin returns to Moscow], Stockholm: SR, TV1, April 14, 1961.

2. Nicholas Barnett, "'RUSSIA WINS SPACE RACE': The British Press and the Sputnik Moment, 1957," *Media History* 19: 2 (2013), 182–195.

3. Lars Lundgren, "Live from Moscow: The Celebration of Yuri Gagarin and Transnational Television in Europe," *VIEW. Journal of European Television History and Culture* 1: 2 (2012), 45–55.

4. We are using the term "global broadcasting" when referring to the idea embraced at the time. When discussing specific broadcasts, such as Gagarin's return to Moscow, we refer to it as "transnational," given that it was far from global in its reach.

5. George Du Maurier, "Edison's Telephonoscope (Transmits Light as Well as Sound)', Almanac for 1879," *Punch* 75 (December 9, 1878).

6. William Uricchio, "Storage, Simultaneity, and the Media Technologies of Modernity," in John Fullerton and Jan Olsson (eds.), *Allegories of Communication: Intermedial Concerns from Cinema to the Digital* (Bloomington: Indiana University Press, 2004), 123–38; William Uricchio, "Television's First Seventy-Five Years: The Interpretive Flexibility of a Medium in Transition," in Robert Kolker (ed.), *Oxford Handbook of Film and Media Studies* (Oxford: Oxford University Press, 2008), 286–305; Brian Winston, *Media, Technology and Society: A History: From the Telegraph to the Internet* (London: Routledge, 1998).

7. "Reading the Stars à la Mode," *Punch* 103 (August 20, 1892), 78.

8. Joshua Nall, *News from Mars: Mass Media and the Forging of a New Astronomy, 1860–1910* (Pittsburgh: University of Pittsburgh Press, 2019), 95–96.

9. See Lisa Parks, *Cultures in Orbit: Satellites and the Televisual* (Durham, NC: Duke University Press, 2005), chapter 2.

10. On the concept of ideation, see Winston, *Media, Technology and Society*.

11. For a history of undersea cables for telegraphy and telephony, see Nicole Starosielski, *The Undersea Network* (Durham, NC: Duke University Press, 2015); Hugh R. Slotten, *Beyond Sputnik and the Space Race: The Origins of Global Satellite Communications* (Baltimore: Johns Hopkins University Press, 2022).

12. James Schwoch, *Global TV: New Media and the Cold War, 1946–69* (Urbana: University of Illinois Press, 2009), 11.

13. Herta Herzog, "Radio—The First Post-War Year," *Public Opinion Quarterly* 10: 3 (1946), 297–313; James C. Foust, "The 'Atomic Bomb' of Broadcasting: Westinghouse's

'Stratovision' Experiment, 1944–1949," *Journal of Broadcasting & Electronic Media* 55: 4 (2011), 510–525.

14. Sarnoff to Truman, November 30, 1948; Sarnoff, "Outline of Proposal," November 30, 1948; both in Truman Presidential Papers, WHCF: Confidential File, State Department Correspondence File, 1948–49, Box 39, Folder "State Department, Correspondence, 1948–49 [5 of 6], Truman Presidential Library: cited in James Schwoch, "Crypto-Convergence, Media, and the Cold War: The Early Globalization of Television Networks in the 1950s," *Media in Transition Conference*, Massachusetts Institute of Technology (MIT), Cambridge, MA, May 2002.

15. Jennifer S. Light, "Facsimile: A Forgotten 'New Medium' from the 20th Century," *New Media & Society* 8: 3 (2006), 355–78, 365.

16. Schwoch, *Global TV*, 88; see also Schwoch, "Crypto-Convergence," 2002.

17. NARCOM was to employ so-called tropospheric scatter to get the UHF signal across. The method exploits a phenomenon in which radio waves are scattered as they hit the troposphere, allowing a concentrated and very narrow beam to be picked up by large, paraboloid antenna.

18. For an account of the idea of radio broadcasting as a means of international peace and understanding, see Simon J. Potter, *Wireless Internationalism and Distant Listening: Britain, Propaganda, and the Invention of Global Radio, 1920–1939* (Oxford: Oxford University Press, 2020).

19. Dana Mustata, "Geographies of Power: The Case of Foreign Broadcasting in Dictatorial Romania," in Alexander Badenoch, Andreas Fickers, and Christian Henrich-Franke (eds.), *Airy Curtains in the European Ether: Broadcasting and the Cold War* (Baden-Baden: Nomos Verlag, 2013), 149–174.

20. Lundgren, "Live from Moscow"; Lars Lundgren, "Transnational Television in Europe: Cold War Competition and Cooperation," in Simo Mikkonen and Pia Koivunen (eds.), *Beyond the Divide: Entangled Histories of Cold War Europe* (New York: Peter Lang, 2015), 237–256.

21. Daniel Dayan and Elihu Katz, *Media Events: The Live Broadcasting of History* (Cambridge, MA: Harvard University Press, 1992); Andreas Fickers and Andy O'Dwyer, "Reading between the Lines," *VIEW Journal of European Television History and Culture* 1: 2 (2012), 1–15.

22. For a history of IBU, see Suzanne Lommers, *Europe—On Air: Interwar Projects for Radio Broadcasting* (Amsterdam: Amsterdam University Press, 2012).

23. For a history of ITU, see Gabriele Balbi and Andreas Fickers, *History of the International Telecommunication Union (ITU): Transnational Techno-Diplomacy from the Telegraph to the Internet* (Berlin: De Gruyter, 2020). ITU organized the Space Radio Conference in Geneva in 1963, seeking to establish an international agreement on space radio. For a detailed account of the conference, as well as negotiations over frequency allocations

between the Soviet Union, the US, and other Western countries, see Hugh R. Slotten, "The International Telecommunications Union, Space Radio Communication, and U.S. Cold War Diplomacy, 1957–1663," *Diplomatic History* 37: 2 (2013), 313–371. For an excellent account of the role of international organizations such as UNESCO and ITU in the development of satellite governance, decolonization and the use of telecommunications for mass media development, see Sarah Nelson, "A Dream Deferred: UNESCO, American Expertise, and the Eclipse of Radical News Development in the Early Satellite Age," *Radical History Review*, 141 (2021), 30–59; Sarah Nelson, "Networking Empire: International Organizations, American Power, and the Struggle over Global Communications in the 20th Century" (PhD dissertation, Vanderbilt University, 2021).

24. Between its founding in 1925 and 1939, IBU grew from ten to fifty-nine member-organizations, with the core of its broadcasters being European.

25. Nina Wormbs, "Technology-Dependent Commons: The Example of Frequency Spectrum for Broadcasting in Europe in the 1920s," *International Journal of the Commons* 5: 1 (2011), 92–109.

26. On program exchanges such as "National Nights" and "European Concerts," see Lommers, *Europe—On Air.*

27. Voting rights, as we will see in chapter 3, was a contested issue in the negotiations leading up to the permanent agreement on Intelsat.

28. The EBU and OIR were thus early examples of regional broadcasting organization. The 1960s witnessed the emergence of a number of regional broadcasting organizations such as the Union of National Radio and Television Organizations of Africa (URTNA, 1962), The Asian Broadcasting Union (ABU, 1964), and the Inter-American Broadcast Association (1965). In addition to technical standardization among national broadcasters (e.g., regarding line standards), the aim of these regional organizations was to promote program exchange. Walter B. Emery, *National and International Systems of Broadcasting: Their History, Operation, and Control* (Minneapolis: Michigan State University Press, 1969); Lyombe Eko, "Steps toward Pan-African Exchange: Translation and Distribution of Television Programs across Africa's Linguistic Regions," *Journal of Black Studies* 3: 31 (2001), 365–379.

29. The "T" in "OIRT," which stood for "Television," was added in 1960.

30. "Creation of the Intervision Network," *OIR Information*, No 2, Special edition, February 1960. OIRT Documents Correspondence, 1950–1960, Folder O6, EBU Archive, Geneva, Switzerland.

31. For an account for a series of meetings between the EBU and OIRT, see Lundgren, "Live from Moscow."

32. Kari Ilmonen, "The Basis of It All—Technology," in Rauno Endén (ed.), *Yleisradio 1926–1996: A History of Broadcasting in Finland* (Helsinki: Finnish Historical Society, 1996), 255.

33. In many respects, the Gagarin broadcast was similar to the coronation of Queen Elisabeth II in 1953, which Daniel Dayan and Elihu Katz have argued as a prime example of what they dubbed a "media event." Daniel Dayan and Elihu Katz, *Media Events* (Cambridge, MA: Harvard University Press, 1992).

34. Ernest Eugster, *Television Programming across National Boundaries: The EBU and OIRT Experience* (Dedham, MA: Artech House, 1983), 106.

35. Christian Henrich-Franke and Regina Immel, "Making Holes in the Iron Curtain? The Television Programme Exchange across the Iron Curtain in the 1960s and 1970s," in Alexander Badenoch, Andreas Fickers, and Christian Henrich-Franke (eds.), *Airy Curtains in the European Ether: Broadcasting and the Cold War* (Baden-Baden: Nomos Verlag, 2013), 177–213.

36. Eugster, *Television Programming*. For more recent accounts of program exchanges between Eastern and Western Europe during the Cold War, see Alexander Badenoch, Andreas Fickers, and Christian Henrich-Franke (eds.), *Airy Curtains in the European Ether: Broadcasting and the Cold War* (Baden-Baden: Nomos Verlag, 2013); Thomas Beutelschmidt, *Ost–West–Global: Das Sozialistische Fernsehen im Kalten Krieg* (Leipzig: Vistas, 2017); Lundgren, "Transnational Television in Europe." For an overview of global television flows in the early 1970s, see Kaarle Nordenstreng and Tapio Varis, "Television Traffic: A One-Way Street? A Survey and Analysis of the International Flow of Television Programme Material," *UNESCO Reports and Papers on Mass Communication*, 1974; Tapio Varis, "Global Traffic in Television," *Journal of Communication* 24: 1 (1974), 102–109.

37. The contemporary BBC documents simply state Aston Villa versus Soviet Union, but the home team was Dynamo Moscow, winning 2–0. Director of Television [Kenneth Adam] to Director General [Hugh Carleton Greene], "Kharlamov Visit: Television Service and Russia," July 9, 1963, *Cultural Relations with Russ, Jan 1963–Dec 1963*, E2/719/7, BBC Written Archives Centre, 2. Again, the contemporary BBC documents recorded the wrong title, "News from the Zoo from Moscow Zoo." Kenneth Adam to Hugh Carleton Greene, "Kharlamov Visit: Television Service and Russia," July 9, 1963, E2/719/7, *Cultural Relations with Russ, Jan 1963–Dec 1963*, E2/719/7, BBC WAC, 2. Thanks to the BBC Genome Project, it is now possible to access the titles as they were presented to the British viewers in *Radio Times*, https://genome.ch.bbc.co .uk/schedules/bbctv/1961-10-04 (accessed November 12, 2020).

38. David Attenborough to H. T. Tel., "Subject: Zoo Quest," February 18, 1958, *Zoo Quest—Russia Suggestion*, T6/443/1, BBC WAC.

39. Leonard Miall to Dmitrii Chestnokov, March 21, 1958, *Zoo Quest—Russia Suggestion*, T6/443/1, BBC WAC, 2.

40. Vasily Evgenyev to Leonard Miall, April 25, 1958, *Zoo Quest—Russia Suggestion*, T6/443/1, BBC WAC.

41. BBC staff involved in both *Zoo Quest* and later exchanges and programming included Leonard Miall, assistant controller of current affairs, Kenneth Adam, director of television, and J. B. Clark, director of external broadcasting.

42. Lars Lundgren, "(Un)familiar Spaces of Television Production: The BBC's Visit to the Soviet Union in 1956," *Historical Journal of Film, Radio and Television* 37: 2 (2017), 315–332.

43. Sonja de Leeuw and Dana Mustata have demonstrated the role of individuals in a transnational television landscape; see Sonja de Leeuw, "Transnationality in Dutch (Pre) Television," *Media History* 16: 1 (2010), 13–29; Dana Mustata, "Architecture Matters: Doing Television History at Ground Zero," *Journal of Popular Television* 7: 2 (2019), 177–199.

44. "Report of the Meeting of Technical Delegates: EBU-OIRT Meeting, Geneva 3rd to 6th February 1960, Final Resolution," February 6, 1960, Appendix 2, O.A./1446-Com.T./25-Com.Pro./377-Com.J./295, EBU.

45. Michele Hilmes, *Network Nations: A Transnational History of British and American Broadcasting* (London: Routledge, 2012).

46. Bell Telephone System, "Project Telstar," n.d.

47. Peter Redfield, *Space in the Tropics: From Convicts to Rockets in French Guiana* (Berkeley: University of California Press, 2000), 144.

48. The idea of using satellites in geosynchronous orbit for global broadcasting is usually attributed to Arthur C. Clarke. See John M. Logsdon, Roger D. Launius, David H. Onkst, and Stephen J. Garber, *Exploring the Unknown: Selected Documents in the History of the US Civilian Space Program. Volume 3; Using Space* (Washington, DC: NASA, 1998). Arthur C. Clarke, "The Space-Station: Its Radio Applications," in John M. Logsdon, Roger D. Launius, David H. Onkst, and Stephen J. Garber (eds.), *Exploring the Unknown* (Washington, DC: NASA, [1945]1998), 12–15; Arthur C. Clarke, "Extra-Terrestrial Relays: Can Rocket Stations Give World-wide Radio Coverage?" *Wireless World* 51: 10 (1945), 305–308.

49. This was demonstrated by Arthur C. Clarke in his famous essay in which he predicted satellite communications; Clarke, "Extra-Terrestrial Relays."

50. Edward Jones-Imhotep, *The Unreliable Nation: Hostile Nature and Technological Failure in the Cold War* (Cambridge, MA: MIT Press, 2017).

51. Much of the following information about the Soviet satellite system is based on the memoirs of Boris Chertok. It should be noted that Chertok's work has earned a highly privileged position as the go-to source on the history of the Soviet space program. However, as pointed out by Slava Gerovitch, this master narrative may well be complicated, and more voices should be heard. Gerovich's interviews with military personnel, space engineers, and cosmonauts provide such complimentary, and sometimes contrasting,

perspectives and descriptions of the Soviet space program, although mainly concerned with human spaceflight. Boris Chertok, *Rockets and People. Vol. III: Hot Days of the Cold War* (Washington, DC: NASA, 2009); Slava Gerovitch, *Voices of the Soviet Space Program: Cosmonauts, Soldiers, and Engineers Who Took the USSR into Space* (Basingstoke, UK: Palgrave Macmillan, 2014); Slava Gerovitch, *Soviet Space Mythologies: Public Images, Private Memories, and the Making of a Cultural Identity* (Pittsburgh: University of Pittsburgh Press, 2015).

52. Chertok, *Rockets and People*, 462. The US launch site at Cape Canaveral was in this respect at a great advantage, not least due to its location near the equator, which allowed a significantly less difficult launch. Later, European satellites would be launched not from Europe, but from French Guiana, from which the geostationary orbit more easily could be accessed. Redfield, *Space in the Tropics*.

53. Chertok, *Rockets and People*, 463.

54. Chief Engineer, Television to C. P. Tel. et al, "Russian Satellites Molniya I and II," Aug 24, 1966, *Foreign Space Communication, Space Communication—General, File 2c, 1965–1966*, T38/25/6, BBC WAC.

55. For more on Telstar, see Schwoch, *Global TV*; James Schwoch, "Removing Some Sense of Romantic Aura of Distance and Throwing Merciless Light on the Weaknesses of the American Life': Transatlantic Tensions of Telstar, 1961–1963," in Alexander Badenoch, Andreas Fickers, and Christian Henrich-Franke (eds.), *Airy Curtains in the European Ether: Broadcasting and the Cold War* (Baden-Baden: Nomos Verlag, 2013): 271–294; Slotten, *Beyond Sputnik*; David J. Whalen, *The Rise and Fall of COMSAT, Technology, Business, and Government in Satellite Communications* (Basingstoke: Palgrave Macmillan, 2014). During its path into orbit, Telstar was exposed to extremely high levels of radiation as a result of a high-altitude nuclear test known as Project Starfish Prime, with a 1.45 megaton explosion detonated on July 9, 1962, the day before Telstar's launch. The radiation significantly shortened the life span of Telstar and as a consequence "[t]he greatest American device yet developed for global communication fell victim to the greatest American device yet developed for global destruction." See James Schwoch, "The Curious Life of Telstar: Satellite Geographies from 10 July 1962 to 21 February 1963," in Jörg Döring and Tristan Thielmann (eds.), *Mediengeographie: Theorie—Analyse—Diskussion* (Bielefeld: transcript Verlag 2009), 342f.

56. Schwoch, "'Removing Some Sense of Romantic Aura'," 276.

57. Cf. Slotten, *Beyond Sputnik*; Whalen, *The Rise and Fall of COMSAT*.

58. INTELSAT IV, launched by the end of December 1965, allowed 6,000 voice circuits. See Whalen, *The Rise and Fall*, 61.

59. On space debris, see Michael Clormann and Nina Klimburg-Witjes, "Troubled Orbits and Earthly Concerns: Space Debris as a Boundary Infrastructure," *Science, Technology, & Human Values* 47: 5 (2021): 960–985; Alice Gorman, *Dr. Space Junk vs. the Universe: Archaeology and the Future* (Cambridge, MA: MIT Press, 2019), ch. 4; Lisa

Ruth Rand, "Falling Cosmos: Nuclear Reentry and the Environmental History of Earth Orbit," *Environmental History* 24:1 (2018): 78–103.

60. "First Phone Call from Space," https://www.youtube.com/watch?v=NSA8scez6Is, accessed March 24, 2023.

61. These showcases were not restricted to broadcasting in any way. During its brief life span, Telstar carried a number of such ceremonial and celebratory telephone calls, such as by linking researchers in Finland and the US by phone in July 1962. Schwoch, "'Removing Some Sense of Romantic Aura'," 288f.

62. The British also describes the French satellite station at Pleumeur-Bodou as a mere replica of the US station in Andover, Maine, and also one being constructed at a much higher cost than the Earth station in Goonhilly. "Notes from Goonhilly Ground Station," July 11, 1962, TCB 2/184, BT Archives, British Telecom, London; from General Post Office to Prime Minister, "Telstar," July 1962, TCB 2/184, BT Archives, British Telecom, London. We would like to thank James Schwoch for drawing our attention to this source.

63. Schwoch, "'Removing Some Sense of Romantic Aura',"; Wolfgang Degenhardt and Elisabeth Strautz, *Auf der Suche Nach dem Europäischen Programm: Die Eurovision 1954–1970* (Baden-Baden: Nomos Verlag, 1999).

64. Schwoch, "'Removing Some Sense of Romantic Aura',"; Degenhart and Strautz, *Auf der Suche*. Experimental broadcasts such as "America to Europe" were not new to broadcasting. We have already mentioned the Calais experiments and the so-called Paris week as early experiments in transnational television broadcasting. In January 1926, International Radio Week was organized, with US radio stations broadcasting to listeners across the Atlantic. See Potter, *Wireless Internationalism*, 29.

65. "Telstar, Europe to America," https://www.youtube.com/watch?v=ponlADg8DbU, accessed March 24, 2023.

66. The plans to include Soviet Central Television had been in the making for some time, and in mid-May 1962, Aubrey Singer got a message stating that if Soviet Television would be presented with a formal invitation, its reply would be favorable. Joanna Spicer, "EBU Satellite Project: Note for File," May 17, 1962, T38/4, BBC WAC.

67. Aubrey Singer to Konstantin Kuzakov, June 3, 1962, *Foreign Space Communication, Europe to America, Countries—Poland and Russia 1962*, T38/4, BBC WAC.

68. "Script for Moscow Sequence," n.d., *Foreign Space Communication, Europe to America, Countries—Poland and Russia 1962*, T38/4, BBC WAC.

69. Robert M. Evans to Edward W. Ploman, June 1, 1962, *Foreign Space Communication, Europe to America, Countries—Poland and Russia 1962*, T38/4, BBC WAC.

70. This may explain why, during a press conference in New York City on July 21, 1962, Aubrey Singer was reluctant to disclose any information about the content of the "Europe to America" broadcast. Schwoch, "'Removing Some Sense of Romantic Aura'," 280.

71. Konstantin Kuzakov to Aubrey Singer, June 27, 1962, *Foreign Space Communication, Europe to America, Countries—Poland and Russia 1962*, T38/4, BBC WAC.

72. Aubrey Singer to Mayorov, June 28, 1962, *Foreign Space Communication, Europe to America, Countries—Poland and Russia 1962*, T38/4, BBC WAC.

73. Olof Rydbeck to Michael Harmolov, July 4, 1962, *Foreign Space Communication, Europe to America, Countries—Poland and Russia 1962*, T38/4, BBC WAC.

74. Aubrey Singer, n.d., *Foreign Space Communication, Europe to America, Countries— Poland and Russia 1962*, T38/4, BBC WAC.

75. Chertok, *Rockets and People*, 453.

76. Another example, described by Boris Chertok, was the broadcast of the funeral of Sergey Korolev, the chief designer of the Soviet space program, in January 1966: "Korolev's funeral was successfully broadcast via Molniya-1 [no. 4] to Vladivostok. One month after Korolev's funeral, the satellite ceased to operate." Chertok, *Rockets and People*, 512. For more on the centrality of calendrical holidays to Soviet television broadcasting, see Christine Evans, *Between Truth and Time. A History of Soviet Central Television* (New Haven, CT: Yale University Press, 2016), 82–114.

77. Chertok, *Rockets and People*, 510. On the Franco-Soviet cooperation regarding color television, see Andreas Fickers, "The Techno-Politics of Colour Britain and the European Struggle for a Colour Television Standard," *Journal of British Cinema and Television* 7: 1 (2010), 95–114.

78. "American Embassy Paris to Secretary of State, Subject: Franco-Soviet Television Relay by Satellite," Airgram A-1129, December 11, 1965, file Tel 6, Subject-Numeric File, Record Group 59: General Records of the Department of State, National Archives at College Park. For more on the Franco-Soviet cooperation, see Isabelle Gourne, "De Passer les Tensions Est-Ouest pour la Conquête de l'Espace: La Coopération Franco-Soviétique au Temps de la Guerre Froide." *Cahiers SIRICE* 2 (16): 49–67; Andrew Jenks, Securitization and Secrecy in the Late Cold War: The View from Space." *Kritika: Explorations in Russian and Eurasian History* 21: 3 (2020): 659–689.

79. American Embassy Paris to Secretary of State, "Pass Commerce," October 6, 1966, Tel 6 1964–1966 SNF, RG59, NACP.

80. Cf. Slotten, *Beyond Sputnik*, 66f.

81. Chertok, *Rockets and People, Vol. III*, 466.

82. Later, the Intersputnik system would expand and carry not only domestic telephone communications, but also link countries such as Cuba and Nicaragua for international telephony.

83. See chapter 2 for more discussion of the challenges of ground station construction in the Soviet Far East and Far North.

84. A. Varbansky, "The Development of the Broadcast Television Network in the USSR: A Translation," *Journal of the SMPTE* 83: 11 (1974), 897–900.

85. Nikolai Mesiatsev, *Gorizonty i vertikaly moei zhizni* (Moscow: Vagrius Press, 2005), 509–511.

86. Mesiatsev, *Gorizonty i vertikaly moei zhizni*, 512.

87. Of course, geostationary broadcasting was far from being Korolev's brainchild in 1964; Mesiatsev too was presumably familiar with the basic premise of geostationary satellite communications, based on the original proposal by Arthur Clark, whose work was well known in the Soviet Union.

88. Geoffrey C. Bowker and Susan Leigh Star, *Sorting Things Out: Classification and Its Consequences* (Cambridge, MA: MIT Press, 2000), 34.

89. Bell Telephone System, "Project Telstar." For example, Telstar almost immediately experienced problems due to the nuclear radiation of Project Starfish. Soviet satellites faced similar obstacles to remaining in service. Schwoch, "The Curious Life of Telstar," 342f; Chertok, *Rockets and People*, 483, 510f.

CHAPTER 2

1. James Farry and David A. Kirby, "The Universe Will Be Televised: Space, Science, Satellites and British Television Production, 1946–1969," *History and Technology* 28: 3 (2012), 311–333. On socialist television as public service broadcasting, see Anikó Imre, *TV Socialism* (Durham, NC: Duke University Press, 2016).

2. Lisa Parks has written by far the most detailed analysis of "Our World," describing it as a "satellite spectacular" designed to offer viewers the experience of "global presence." However, this sense of belonging to a world of satellite technology-enabled modernity, she argues, was limited to those countries and people able to receive and participate in the broadcast, which were chiefly in the wealthier Global North. Parks argues that the global presence represented in "Our World" should thus be seen as a Western fantasy, rooted in discourses of Western modernization and exposing "neocolonial strategies at work." Lisa Parks, *Cultures in Orbit: Satellites and the Televisual* (Durham, NC: Duke University Press), 23, 29.

3. Robert Bird, "Revolutionary Synchrony: A Day of the World," *Baltic Worlds* 3, 2017, 45–52. See also Sabina Mihelj and Simon Huxtable, "The Challenge of Flow: State Socialist Television between Revolutionary Time and Everyday Life," *Media, Culture & Society* 38: 3 (2015), 332–348.

4. Yuri Fokin, "Kak vse eto nachinalos [How it All Began]," *Sovietskoe radio I televidenie [Soviet Radio and Television]*, 2 (10) (1968), 10.

5. Dina Fainberg, *Cold War Correspondents: Soviet and American Reporters on the Ideological Frontlines* (Baltimore: Johns Hopkins University Press, 2021); Thomas Wolfe, *Governing Soviet Journalism: The Press and the Socialist Person after Stalin* (Bloomington: Indiana University Press, 2005); Vladislav Zubok, *Zhivago's Children: The Last Russian Intelligentsia* (Cambridge, MA: Harvard University Press, 2009); Simon Huxtable,

News from Moscow: Soviet Journalism and the Limits of Postwar Reform (Oxford: Oxford University Press, 2022).

6. Stephen Lovell, *Russia in the Microphone Age*, (Oxford: Oxford University Press, 2015), 77; see also Wolfe, *Governing Soviet Journalism*; Bird, "Revolutionary Synchrony."

7. See, for example, Oksana Sarkisova, *Screening Soviet Nationalities: Kulturfilms from the Far North to Central Asia* (London: I. B Tauris, 2017).

8. In the end, fourteen countries participated in the production of "Our World." After the Soviet Union and four other East European countries left the project, Denmark was added.

9. "Our World" has garnered some academic interest as an iconic and groundbreaking broadcast, whereas "One Hour in the Life of the Motherland" has remained unknown and forgotten even in post-Soviet Russia, beyond a couple of brief mentions in the Russian-language memoirs of Soviet television producers.

10. See Kristin Roth-Ey, *Moscow Prime Time: How the Soviet Union Built the Media Empire That Lost the Cultural Cold War* (Ithaca, NY: Cornell University Press, 2011); Sarkisova, *Screening Soviet Nationalities*.

11. Aubrey Singer, "Promotional Material—Programme Story, Note by Aubrey Singer," n.d., T14/2723/1, BBC Written Archives Centre, 1.

12. Jerôme Bourdon, "Live Television Is Still Alive: On Television as an Unfulfilled Promise," *Media, Culture & Society* 22: 5 (2000), 531–556; Philip Auslander, *Liveness: Performance in a Mediatized Culture* (London: Routledge, 2008); Paddy Scannell, *Television and the Meaning of "Live,"* (Cambridge, UK: Polity, 2014).

13. The idea that time and space are inseparably linked in any artistic or other representational project is well established in, for example, literary criticism and philosophy. Geographers have similarly called for the reintegration of time into the humanistic and social scientific study of space. In media research, the problem of time and space has long been central, often inspired by the work of Harold A. Innis, James Carey and others. Mihkail M. Bahktin, "Forms of Time and of the Chronotope in the Novel: Notes toward a Historical Poetics," in Brian Richardson (ed.), *Narrative Dynamics. Essays on Time, Plot, Closure, and Frames* (Columbus: Ohio State University Press, 2002), 15–16; Henri Lefebvre, *The Production of Space* (Oxford: Blackwell, 1974); David Harvey, *The Condition of Postmodernity: An Enquiry into the Origin of Cultural Change* (Oxford: Blackwell, 1989); Doreen Massey, "Space-Time, 'Science' and the Relationship between Physical Geography and Human Geography," *Transactions of the Institute of British Geographers* 24 (1999), 261–276; Harold A. Innis, *The Bias of Communication* (Toronto: University of Toronto Press, 1951); James W. Carey, *Communication as Culture: Essays on Media and Society* (Boston: Unwin Hyman, 1989).

14. See Christine Evans and Lars Lundgren, "Geographies of Liveness: Time, Space, and Satellite Networks as Infrastructures of Live Television in the Our World Broadcast," *International Journal of Communication* 10 (2016), 5362–5380.

15. Lisa Parks refers to the notions of "scheduled liveness" and "time zooning" as means to articulate global presence. To construct global presence, "Our World" had to negotiate preexisting temporal structures (i.e., impose a temporal regime upon national broadcasters). "Time zooning" refers to the practice of articulating the multiple temporalities at play during satellite spectaculars such as "Our World." Parks, *Cultures in Orbit*, 7–42; Auslander, *Liveness*, 3.

16. Asif A. Siddiqi, "Cosmic Contradictions: Popular Enthusiasm and Secrecy in the Soviet Space Program," in James T. Andrews and Asif A. Siddiqi (eds.), *Into the Cosmos: Space Exploration and Soviet Culture* (Pittsburgh: University of Pittsburgh, 2011), 47–76.

17. "Aubrey Singer to Henry Trofimenko," December 9, 1965, *Our World Correspondence*, T14/2722/2, BBC WAC.

18. Vanessa Ogle has pointed out that the word "world" was frequently used to signal interconnectedness during the first decades of the twentieth century, in such expressions as "world calendars" or "world languages" such as Esperanto. Vanessa Ogle, *The Global Transformation of Time, 1870–1950* (Cambridge, MA: Harvard University Press, 2015), 97.

19. Aubrey Singer, "Russian Attitude to 'Round the World in 80 Minutes'," March 17, 1966, *Our World Correspondence*, T14/2722/2, BBC WAC.

20. "Aubrey Singer to Anatolii Bogomolov," March 7, 1966, *Our World Correspondence*, T14/2722/2, BBC WAC.

21. It turned out that 1966 proved too soon for both practical reasons (the fact that Soviet satellite infrastructure was not yet complete) and political ones; the Soviet side had a number of concerns about the content of the broadcast. Singer, "Russian Attitude to 'Round the World in 80 Minutes.'"

22. Singer, "Promotional Material."

23. We suggest that this hierarchy was devised according to the show's producers' own values, which are of course not universal; from other cultural perspectives, there is nothing inherently shameful or dehumanizing about making the birthing human body visible.

24. The full broadcast of "Our World" is currently available on YouTube: https://www.youtube.com/watch?v=s3LmQFt4pQc, accessed March 24, 2023. See also Parks's account of this segment in *Cultures in Orbit*, 29.

25. "Spravka "o khode proektirovaniie i stroitel'stva ob"ektov kosmicheskoi sviazi po sostoianniu na 15.03.68 g." [Report "On progress in the planning and construction of space communications assets as of March 15, 1968] (March 15, 1968). RGAE F. 3527, op. 55, d. 20, l. 5.

26. "Letter from Aubrey Singer to Tony Jay," March 16, 1967, *Our World General*, T14/2723/3, BBC WAC, 2.

27. "Letter from Aubrey Singer to Tony Jay," 3.

28. "Maryla Wisniewska to Aubrey Singer," April 18, 1967, *Our World Correspondence*, T14/2722/2, BBC WAC, 1.

29. Mihelj and Huxtable, "The Challenge of Flow," 334; Sabina Mihelj and Simon Huxtable, *From Media Systems to Media Cultures: Understanding Socialist Television* (Cambridge: Cambridge University Press, 2018).

30. "Maryla Wisniewska to Aubrey Singer," 1.

31. Aubrey Singer and Andrew Wilson, "Record of a Meeting with Mr. Ivanov, Vice Chairman of the State Committee of Radio and Television in Moscow. He Is Also the Director of Television," July 21, 1966, *Our World Correspondence*, T14/2722/2, BBC WAC.

32. "Record of Meeting with Mr. Ivanov on 22nd July at Moscow Television," n.d., *Our World Technical*, T14/2723/2, BBC WAC, 2.

33. "Special Notice No. T.3/67 Our World," April 24, 1967, *Our World Technical*, T14/2739/1, BBC WAC, 2.

34. "Our World Network Map, Conferences and Control Networks," *Our World Technical*, T14/2739/2, BBC WAC.

35. Hugh Carleton Greene, "Draft Article for Radio Times," May 26, 1967, *Our World 1966–1967*, E2/864/1, BBC WAC, 2.

36. Nicole Starosielski, *The Undersea Network* (Durham, NC: Duke University Press, 2015).

37. "Our World," https://www.youtube.com/watch?v=s3LmQFt4pQc.

38. Greene, "Draft Article for Radio Times," 2. See also Jane Feuer, "The Concept of Live Television: Ontology as Ideology," in E. Ann Kaplan (ed.), *Regarding Television: Critical Approaches—An Anthology* (Los Angeles: American Film Institute, 1983).

39. Singer, "Promotional Material," 1.

40. Greene, "Draft Article for Radio Times," 2.

41. Ogle, *The Global Transformation of Time*, 206.

42. Andy Wiseman, "Translation of Telex from G. Ivanov to A. Singer," June 21, 1967, *Our World General*, T14/2723/3, BBC WAC.

43. "Glavnoe Upravleniie Kosmicheskoi Sviazi 1967–1970 gg." [Main Directorate for Space Communications] (1967–70). F. 3527 "Ministerstvo Sviazi SSSR" [Ministry of Communications USSR]. op.55, d. 5–17, 25–32, 37–38, 43–48. Russian State Economic Archive (Rossiiskii Gosudarstvennyi Arkhiv Ekonomiki, Moscow [RGAE]).

44. For example, the head of the Space Department (*Kosmicheskii otdel*) of the Ministry of Communications expressed frustration that sites for Molniya Earth stations were being selected before geological testing was complete, leading to problems with frozen soil and ice in the construction process. Letter from Vladimir Minashin to

Nikolai Talyzin, June 29, 1971, "Dokladnye zapiski i spravki Upravleniia rukovod-stvu Min Sviazi SSSR o deiatel'nosti Upravleniia," RGAE F. 3527, op. 63, d. 778, l. 11. For more on the challenges of defining and predicting the behavior of permafrost, see Pey-Yi Chu, *The Life of Permafrost: A History of Frozen Earth in Russian and Soviet Science* (Toronto: University of Toronto Press, 2020).

45. "Spravka "o khode proektirovaniie i stroitel'stva ob"ektov kosmicheskoi sviazi po sostoianniu na 15.03.68 g." [Report "On progress in the planning and construc-tion of space communications assets as of March 15, 1968] (1968, March 15). F. 3527, op. 55, d. 20, l. 5. RGAE.

46. John Downing, "The Intersputnik System and Soviet Television," *Soviet Studies* 37: 4 (1985), 468.

47. Parks, *Cultures in Orbit*, 42.

48. "Protokol # 25 zasedaniia komiteta po radioveshchaniiu i televideniiu pri SM SSSR. "O radio i televizionnykh programmakh v dni prazdnovaniia 50-oi godovsh-chiny velikoi Oktiabr'skoi sotsialisticheskoi revoliutsii." October 2, 1967. Moscow, GARF. F. 6903, op. 1 d. 941, ll. 85–86.

49. See Nikolai Mesiastev's mention of the broadcast in his memoir, such as *Gori-zonty i vertikaly*, 489–490. A review of the broadcast was published several months later in the main professional journal for television workers, see Ia Damskii, "Minuta v efire. Korotkaia retsenziia [A Minute on Air]," *Sovietskoe Radio i Televidenie*, no. 2 (February 1968), 10–14.

50. Christine E. Evans, "A 'Panorama of Time': The Chronotopics of Programma 'Vremia'," *Ab Imperio* 2, 2010, 121–146.

51. Sarkisova, *Screening Soviet Nationalities*.

52. On "One Day in the World," see Bird, "Revolutionary Synchrony."

53. In both "On Day in the World" and "One Hour in the Life of the Motherland," the childbirth motif was presented optimistically, with each child offering a glimpse of a communist future. "Our World," by contrast, was more ambiguous, linking the image of a future generation to an ominous shrinking of resources and space itself. Aleksei Adzhubei and L. P. Grachev (eds.), *Odin den' mira: Sobytiia 27 sentiabria 1960 g.* (Moscow: Izvestiia, 1960).

54. Gennadii Sorokin, "Oruzhie ne dlia kustarei," *Zhurnalist*, no. 1 (January 1968), 31.

55. Sorokin, "Oruzhie ne dlia kustarei."

56. Sorokin, "Oruzhie ne dlia kustarei."

57. The records of the Tbilisi television studio suggest that one of its main activities in 1967 was the production of documentary television programming about Georgia for exchange with other SSRs, for Soviet Central Television in Moscow, for Soviet participation in international expositions like Expo-67 in Montreal, and for socialist

bloc countries such as the German Democratic Republic (GDR). See, for example, National Archive of Contemporary History, Tbilisi f. 1978, Tbilisi Television Studio, op. 1, d. 1303, 1415.

58. "Noble Wilson to Peter Pockley," November 7, 1967, *Our World General*, T14/2723/3, BBC WAC. Soviet critics also noted that the Tashkent segment was of lower quality than the rest of the broadcast, although they emphasized, perhaps euphemistically, its failure to fit in with the style of the rest of the broadcast. Damskii, "Minuta v efire. Korotkaia retsenziia,." 13.

59. Katerina Clark, *Moscow, the Fourth Rome: Stalinism, Cosmopolitanism, and the Evolution of Soviet Culture, 1931–1941* (Cambridge, MA: Harvard University Press, 2011).

CHAPTER 3

1. "The Soviet Statsionar Satellite Communications System: Implications for INTEL-SAT," Interagency Intelligence Memorandum, NIO IMM 76–016, April 1976, CIA: 1. https://www.cia.gov/readingroom/docs/DOC_0000283805.pdf (accessed April 30, 2022).

2. For example, see Odd Arne Westad, *The Cold War: A World History* (New York: Basic Books, 2017). Resistance to this approach has come from national and comparative studies of Cold War technology, as well as a small but growing group of transnational histories of Cold War technology. See, for example, Benjamin Peters, *How Not to Network a Nation: The Uneasy History of the Soviet Internet* (Cambridge, MA: MIT Press, 2016); Eden Medina, *Cybernetic Revolutionaries: Technology and Politics in Allende's Chile* (Cambridge, MA: MIT Press, 2014); Egle Rindzevicute, *The Power of Systems: How Policy Science Opened up the Cold War World* (Ithaca, NY: Cornell University Press, 2016); Ksenia Tatarchenko, *"A House with the Window to the West": The Akademgorosk Computer Center (1958–1993)*, (PhD dissertation, Princeton University, Princeton, NJ, 2013).

3. James Schwoch, *Global TV: New Media and the Cold War, 1946–69* (Urbana: University of Illinois Press, 2009), 153.

4. See Lisa Parks and James Schwoch, *Down to Earth: Satellite Technologies, Industries, and Cultures* (New Brunswick, NJ: Rutgers University Press, 2012); Lisa Parks, *Cultures in Orbit: Satellites and the Televisual* (Durham, NC: Duke University Press, 2005); Lisa Parks, *Rethinking Media Coverage: Vertical Mediation and the War on Terror* (London: Routledge, 2018); Andrew J. Butrica, ed., *Beyond the Ionosphere: Fifty Years of Satellite Communication* (NASA SP-4217, 1997); Ingrid Volkmer, "Satellite Cultures in Europe: Between National Spheres and a Globalized Space," *Global Media and Communication* 4: 3 (2008), 231–244; Hugh R. Slotten, "Satellite Communications, Globalization and the Cold War," *Technology and Culture* 43: 2 (2002); Hugh R. Slotten, *Beyond Sputnik and the Space Race: The Origins of Global Satellite Communications* (Baltimore: Johns Hopkins University Press, 2022).

5. The Plenipotentiary Conference to Establish Definitive Arrangement for the International Telecommunication Satellite Consortium began in February 1969, and the final treaty was signed in August 1971. Until the treaty was signed Intelsat was governed by the interim agreement of 1964.

6. For some, this constituted a betrayal of communications satellites' *technologic imperative*, a form of technological determinism that saw new technologies as having inherent, unalterable qualities. See, for example, Delbert D. Smith, *Communication via Satellite* (Leiden: A. W. Sijthoff, 1976). These commentators admitted that the failure to create a single global satellite network was unsurprising, given the very divergent perspectives even within Intelsat's own membership, not to mention between the US and the Soviet Union. Edward McWhinney, "Review: Communication via Satellite: A Vision in Retrospect, by Delbert D. Smith," *American Journal of International Law* 71: 4 (1977), 834–835.

7. McWhinney, "Review: Communication via Satellite," 835. See also Judith T. Kildow, *INTELSAT: Policy-Maker's Dilemma* (Lexington, MA: Lexington Books, 1973); John Downing, "The Intersputnik System and Soviet Television," *Soviet Studies* 37: 4 (1985), 465–483; Olof Hultén, "The Intelsat System: Some Notes on Television Utilization of Satellite Technology," *International Communication Gazette* 19: 1 (1973), 29–37; Drew McDanie and Lewis A. Day, "INTELSAT and Communist Nations' Policy on Communications Satellites," *Journal of Broadcasting* 18: 3 (1974), 311–322.

8. David Whalen, *The Origins of Satellite Communications, 1945–1965* (Washington, DC: Smithsonian Institution Press, 2002), 81. The Soviet Union felt a similar need to justify space investment by describing the scientific and technical developments for manufacturing initially developed as part of space research. See for example, "Space technology, its influence on science and technology," *Aviatsiia i kosmonavtika*, no. 11 (April 1974), 34.

9. Communication satellites infrastructure followed, on a global scale, the pattern European historians of technology have described as "hidden integration," in which technical networks, such as gas pipelines, to mention only Per Högselius's landmark study, connected European countries in ways that often directly undermined prevailing geopolitical logics. See Thomas J. Misa and Johan Schot, "Inventing Europe: Technology and the Hidden Integration of Europe," *History and Technology* 21: 1 (2005), 1–19, as well as the other articles in this special issue. On gas pipelines as an example of hidden integration, see Per Högselius, *Red Gas: Russia and the Origins of European Energy Dependence* (New York: Palgrave Macmillan, 2013), 2–3. The broader historical literature has also begun to interrogate Cold War binaries. See for example, Simo Mikkonen and Pia Koivunen, *Beyond the Divide: Entangled Histories of Cold War Europe* (New York: Berghahns Books, 2015); Peter Romijn, Giles Scott-Smith, and Joes Segal (eds.), *Divided Dreamworlds? The Cultural Cold War in East and West* (Amsterdam: Amsterdam University Press, 2012). See also the book series *Rethinking the Cold War*, published by de Gruyter.

10. Conference on the Role of International Organizations in the Regulation of Communications in Outer Space, September 21–25, 1969, Talloires, file Economic Tel 6 7/1/69 to 11/1/69, 1967–1969 Subject-Numeric Files, Record Group 59: General Records of the Department of State, National Archives at College Park.

11. For some examples of the ongoing reevaluation of the Soviet relationship to globalization during the Cold War, see Oscar Sanchez-Sibony, "Capitalism's Fellow Traveller: The Soviet Union, Bretton Woods, and the Cold War, 1944–1958," *Comparative Studies in Society and History* 56: 2 (2014), 290–319; Oscar Sanchez-Sibony, *Red Globalization: The Political Economy of the Soviet Cold War from Stalin to Khrushchev*, (Cambridge: Cambridge University Press, 2014); Sari Autio-Sarasmo, "Stagnation or Not? The Brezhnev Leadership and the East-West Interaction," in Dina Fainberg (ed.), *Reconsidering Stagnation in the Brezhnev Era: Ideology and Exchange* (Lanham, MD: Lexington Books, 2016), and Sari Autio-Sarasmo, "Planning in Cold War Europe: Competition, Cooperation, Circulations (1950s–1970s)," in Michael Christian, Sandrine Kott, and Ondrej Matejka (eds.), *Planning in Cold War Europe: Competition, Cooperation, Circulations (1950s–1970s)* (Oldenbourg: De Gruyter Oldenbourg, 2018), 143–164.

12. In this sense, we follow Graham and Marvin's call to analyze infrastructural changes across the developed, developing, and postcommunist worlds, but go further by considering communist-world infrastructures as important actors within the processes of capitalist globalization; Stephen Graham and Simon Marvin, *Splintering Urbanism: Networked Infrastructures, Technological Mobilities and the Urban Condition* (London: Routledge, 2001).

13. Paul N. Edwards, "Meteorology as Infrastructural Globalism," *Osiris* 21: 1 (2006), 230; Paul N. Edwards, *A Vast Machine: Computer Models, Climate Data, and the Politics of Global Warming* (Cambridge, MA: MIT Press, 2010).

14. Graham and Marvin, *Splintering Urbanism*, 33, 35.

15. Nikhil Anand, Akhil Gupta, and Hannah Appel, eds., *The Promise of Infrastructure* (Durham, NC: Duke University Press, 2018). Satellite communication, as part of the larger idea of space exploration and utilization is even further invested in a kind of utopian futurism, see Douglas de Witt Kilgore, *Astrofuturism: Science, Race, and Visions of Utopia in Space* (Philadelphia: University of Pennsylvania Press, 2003).

16. *Communication Satellites: Technical, Economic and International Development*. Staff report prepared for the use of the Committee on Aeronautical and Space Sciences, US Senate. US Government Printing Office, 1962, 46; Kildow, *INTELSAT*, xi.

17. Slotten, *Beyond Sputnik*.

18. Whalen, *The Origins of Satellite Communications*, 81.

19. Kildow, *INTELSAT*, 52–57.

20. The July 8, 1966, Paris edition of the *New York Herald Tribune* quoted Robert Aubinière, director of the Centre Nationale des Etudes Spatiales (CNES), as saying that

"within a few years . . . the Americans will have organized a world-wide network of television by satellite. Unless we by then have national launchers to install an identical French network, even a modest one, our culture will be swept away. Before the end of the century all Africa will speak English." US Embassy Paris to Department of State, Airgram A-206, August 4, 1966, Tel 6, 1964–66 SNF, RG 59, NACP.

21. Downing, "The Intersputnik System," 466.

22. The Franco-Soviet experiments on satellite broadcasting and color television was initiated in 1965 with a series of experimental broadcasts starting in May 1966, see ITU, *Sixth Report by the International Telecommunication Union on Telecommunication and the Peaceful Uses of Outer Space (1967)*, (Geneva: ITU, 1967). For more on Franco-Soviet relations and the SECAM color television system, see Fickers, "The Techno-Politics of Colour".

23. US Embassy Paris to Department of State, Airgram A-1129, December 11, 1965, Tel 6, 1964–66 SNF, RG 59, NACP.

24. Evert Clark, "Propaganda Is Called a Peril of Communications Satellites," *New York Times*, May 5, 1966. On direct satellite broadcasting, see Ithiel de Sola Pool. "Direct Broadcast Satellites and the Integrity of National Cultures," in Kaarle Nordenstreng and Herbert Schiller (eds.), *National Sovereignty and International Communication: A Reader* (Norwood, NJ: Ablex Publishing Corporation, 1979), 120–153; and Kathryn M. Queeney, *Direct Broadcast Satellites and the United Nations* (Alphen aan den Rijn: Sijthoff and Noordhoff, 1978); and Kaarle Nordenstreng and Tapio Varis, "Television Traffic: A One-Way Street? A Survey and Analysis of the International Flow of Television Programme Material," *UNESCO Reports and Papers on Mass Communication*, 1974..

25. Clark, "Propaganda Is Called a Peril."

26. Clark, "Propaganda Is Called a Peril."

27. Clark, "Propaganda Is Called a Peril."

28. In the early1970s, Soviet concerns about direct-to-home satellite broadcasting, which they understood would be controlled by the US, given their significant head start in developing a global system, were very great. These fears also directly mirrored the fears articulated by US officials—Soviet officials worried that the US would distribute the antennae necessary to receive their direct broadcasts for free in targeted countries. See for example, I. V. Vasilieva (Soviet Academy of Sciences, Institute of Applied Social Research), "Sputnikovoe neposredstvennoe veshchaniie, ego problemy i perspektivy [Direct satellite broadcasting: Problems and perspectives]," Report to the Central Committee General Section, February 3, 1971 RGANI F. 5, op. 63, d. 247, l. 23. For more on Soviet fears of DBS, see Kristin Roth-Ey, *Moscow Prime Time: How the Soviet Union Built the Media Empire That Lost the Cultural Cold War* (Ithaca, NY: Cornell University Press, 2011), 220.

29. In March 1962, for instance, Kennedy addressed the issue in a letter to Khrushchev. In addition to such high-level political contacts, US and Soviet delegations

met to discuss the possibility of finding areas of cooperation in space. Khruschev responded to the letter on March 20, 1962, suggesting a number of areas in which the US and the Soviet Union could cooperate in order to further the "peaceful use of outer space," including communications satellites and global television broadcasts. John F. Kennedy to Nikita S. Khruschev, March 7, 1962, Papers of John F. Kennedy. Presidential Papers. White House Staff Files of Pierre Salinger. Subject Files, 1961–1964. Khrushchev/Kennedy Letters, 7 March 1962–20 January 1963. John F. Kennedy Presidential Library and Museum; Nikita S. Khruschev to John F. Kennedy, March 20, 1962, Papers of John F. Kennedy. National Security Files. Carl Kaysen Series, Box 377, John F. Kennedy Presidential Library and Museum; US Embassy Geneva to Department of State, "Space Communication Talks with Soviets," June 16, 1964, TEL 6 Space Communications, TEL 4 1/1/64—TEL 6 4/1/66, 1964–1966 SNF, RG 59, NACP.

30. On the rhetoric of apolitical science during the Cold War, see Audra J. Wolfe, *Freedom's Laboratory: The Cold War Struggle for the Soul of Science* (Baltimore: Johns Hopkins University Press, 2018).

31. Lyndon B. Johnson, "Special Message to the Congress on Communication Policy," August 14, 1967, Gerhard Peters and John T. Wooley, eds, *The American Presidency Project*, http://www.presidency.ucsb.edu/ws/?pid=28390 (accessed January 4, 2018). See also Schwoch, *Global TV*, 150–153.

32. Johnson, "Special Message to the Congress," 1967.

33. US mission to the United Nations to Department of State, New York, Telegram 6920, October 8, 1968, TEL 6, 1967–69 SNF, RG 59, NACP.

34. Interkosmos was founded in November 1965, "Academic B. N. Petrov, Chairman of Interkosmos, to Academic L. Kristanov, Chairman of the Bulgarian Academy of Science's Committee for Study of the Cosmos," September 4, 1967, Arkhiv Rossiiskoi Akademii Nauk (ARAN), Moscow, Russia, Fond 1678 "Sovet po mezhdunarodnomu sotrudnichestvu v oblasti issledovaniia i ispol'zovaniia kosmicheskogo prostranstva "Interkosmos,"" op. 1 d. 9, l. 63. See also Colin Burgess and Burt Vis, *Interkosmos: The Eastern Bloc's Early Space Program* (Chichester: Springer Praxis), 2–3; Thomas Beutelschmidt, *Ost–West–Global: Das Sozialistische Fernsehen im Kalten Krieg* (Leipzig: Vistas, 2017), 277.

35. "Plan mezhdunarodnykh sviazei Sovetskogo Soiuza s sotsialisticheskimi stranami v oblasti issledovaniia i ispol'zovaniia kosmicheskogo prostranstva v mirnykh tseliakh na IV kvartal 1967 g." ARAN f. 1678 "Interkosmos" op. 1, d. 1, l. 152.

36. See Petrov's May 7, 1968 letter to the Chair of the Hungarian Committee for Space Research, urging him to organize a meeting to approve the draft agreement before the August UN meeting in Vienna. ARAN f. 1678 op. 1 d. 34, l. 13.

37. "Zakliuchitel'nyi protokol, g. Budapesht 24–29 iiunia 1968 g." ARAN F. 1678 Sovet "Interkosmos" op. 1, d. 34, l. 91

38. "Ob'iasnitel'naia zapiska," Draft agreement for Mezhdunarodnaia Sistema Sputnikovoi Sviazi, March 1968, ARAN, F. 1678 op. 1 d. 34, l. 24.

39. The connections between the Soviet Intersputnik proposal, as well as, to a lesser extent, the Intelsat interim agreements of 1964, and the basic organizational structure of the United Nations is not surprising, given that these satellite networks were conceived as *both* intergovernmental and commercial agreements. At the same time, both the US and, likely, the Soviet Union sought to avoid subordinating satellite communications agreements to the United Nations Committee on the Peaceful Use of Outer Space (formed in 1959), something that some smaller countries, such as Sweden, had proposed.

40. ARAN f. 1678 op. 1 d. 34, l. 23–24. Understandably, both Intelsat and Intersputnik were interested in recruiting India and Pakistan, given their size and economic power. Both India and Pakistan were among thirteen countries prioritized by the US, meaning (among other things) that the Department of State would support the development of Earth stations in the selected countries. "Memorandum: Communication Satellite Earth Stations Construction in Less Developed Countries," December 31, 1966, Tel 6 1/1/67 to Tel 6 7/1/67, Economic, 1967–1969 SNF, RG 59, NACP.

41. Slotten, "Satellite Communications", 348–349; Slotten, *Beyond Sputnik*.

42. V. A. Racheev, Counselor of the Soviet Embassy to the United States, "Notes from Conversation with Counselor Steiner [Switzerland], 30 Sept. 1968. From the counselor's diary from 25 Sept. 1968." ARAN F. 1678 op. 1 d. 34, l. 142.

43. Racheev, "Notes from Conversation with Counselor Steiner," l. 142–145. Notably, other Swiss officials had taken a different line with the US deputy assistant secretary for transportation and communications, Frank Loy, earlier in the summer, saying that they felt that regional networks should not undermine "logic or economics" of a worldwide system. Frank Loy, "Notes on Talks with Swiss Officials, June 18, 1968," 2, Folder 7/1/68, Economic: TEL 6 7/1/68–11/1/68, 1967–1969 SNF, RG 59, NACP.

44. See, for example, "Otchet Sovetskoi delegatsii o komandirovke vo Frantsiiu / Parizh s 16 po 29 maia 1967 g. na soveshchanie rabochei gruppy No. 2/ kosmicheskaia sviaz'/ sozdannoi v sootvetstvii s protokolom ot 12 oktiabria 1966 goda/ g. Moskva/ o nauchno-tekhnicheskom sotrudnichestve mezhdu SSSR i Frantsiei v oblasti kosmicheskoi sviazi." ARAN f. 1678 op. 1 d. 8 l. 109; and "Otchet o rabote rabochei gruppy No. 2 (kosmicheskaia sviaz') Parizh, 1–11 Oktiabria 1968 g." ARAN F. 1678 Soviet "Interkosmos" op. 1, d. 36, l. 56.

45. "Proekt Soglasheniie ob Uchrezhdenii Mezhdunarodnoi Organizatsii po Ispol'zovaniiu Iskusstvennykh Sputnikov Zemli Dlia Tselei Sviazi," ARAN F. 1678 "Interskosmos," op. 1, d. 9, l. 43–44.

46. This included Intersputnik's status as an independent, commercial entity with "legal personality," allowing it to own its space segment and/or lease space capacity

from other countries on behalf of its members, see "Proekt Soglasheniie ob Uchrezh-
denii Mezhdunarodnoi Organizatsii po Ispol'zovaniiu Iskusstvennykh Sputnikov
Zemli Dlia Tselei Sviazi," ll. 43–44. On European requests for a "legal personality" for
Intelsat, see "European Conference on Satellite Communications. Swiss Contribution.
Working Party on the Definitive Arrangement for INTELSAT. Legal Personality of the
Organization to Succeed INTELSAT," May 7, 1968, Folder 7/1/68, Economic: TEL 6
7/1/68–11/1/68, 1967–1969 SNF, RG 59, NACP.

47. "Memo from US Embassy Moscow to Department of State, re: Intersputnik and
Intelsat," Telegram 5125, August 17, 1968, Folder 8-1-68, Economic: TEL 6 7/1/68–
11/1/68, 1967–1969 SNF, RG 59, NACP.

48. See, for example, Susan Buck, *The Global Commons: An Introduction* (Washington,
DC: Island Press, 2012), 160. James Schwoch, by contrast, notes Soviet openness to
cooperation. Schwoch, *Global TV*, 150–153.

49. See, for example, ARAN F. 1678 Sovet "Interkosmos" op. 1, d. 34, ll. 29–82. Intellectual property that resulted from the networks' work also was to belong to all member-
countries, regardless of who invented it. See "Proekt Soglasheniie ob Uchrezhdenii
Mezhdunarodnoi Organizatsii po Ispol'zovaniiu Iskusstvennykh Sputnikov Zemli Dlia
Tselei Sviazi," ARAN f. 1678 op. 1, d. 9, l. 54.

50. "Zakliuchitel'nyi protokol g. Ulan-Bator 10–17 marta 1970 g." ARAN F. 1678
Sovet "Interkosmos" op. 1, d. 68a, l. 10.

51. US Mission Geneva to Secretary of State, Subject: "Recent Intersputnik Developments," Telegram 2830, August 7, 1969, Economic: Tel 6 7/1/69 to Tel 6 11/1/69,
1967–1969 SNF, RG 59, NACP.

52. See, for example, "Secretary of State to the President, Subject: Soviet INTER-
SPUTNIK Proposal," September 3, 1968, 1, Folder 9-1-68, Economic: TEL 6 7/1/68–
11/1/68, 1967–1969 SNF, RG 59, NACP.

53. "Letter from Academic B. N. Petrov, Chair of Interkosmos, to N. V. Talyzin, Vice
Minister of Communications of the USSR," August 12, 1967, ARAN f. 1678 op. 1, d.
9, l. 61–62.

54. *Foreign Relations of the United States, 1964–1968, Volume XXXIV, Energy Diplomacy
and Global Issues*, ed. Susan K. Holly (Washington, DC: Government Printing Office,
1991) Document 100, https://history.state.gov/historicaldocuments/frus1964-68v34
/d100 (accessed December 15, 2017).

55. U.S. Department of State to All U.S. Diplomatic Posts, Subject: re: 'Intelsat Conference,' 8, Airgram CA-12775, Folder 12-17-68, Economic: TEL 6 11/11/68–1/1/69,
1967–1969 SNF, RG 59, NACP.

56. See, for example, Mongolia's response to the survey, "From Academic B. Shi-
rendyb to Academic Petrov," July 20, 1967, ARAN f. 1678 "Interkosmos" op. 1, d.
9, l. 3.

57. "Mneniie Pol'skoi storony otnositel'no voprosnika po dannym dlia podgotovki predvaritel'nogo proekta MSSS," n.d. (1967), ARAN f. 1678 "Interkosmos," op. 1, d. 9, l. 126.

58. "Zakliuchitel'nyi protokol, g. Budapesht 24–29 iiunia 1968 g." ARAN f. 1678, Sovet "Interkosmos" op. 1, d. 34 l. 98.

59. These overtures were the subject of extensive correspondence between the State Department, COMSAT, and other agencies from 1966–1973. See, for example, "Letter from William K. Miller, Office of Telecommunications, US Department of State to Irving I. Schiffman, Economic Officer, US Embassy Bucharest," September 16, 1968, Tel 6, 1967–1969 SNF, RG 59, NACP.

60. US Embassy Bucharest to US Department of State, Airgram A-12, January 11, 1973, Tel 6, 1970–1973 SNF, RG 59, NACP.

61. Yugoslav officials confidently expected that a Yugoslav Earth station would carry traffic for Romania and Hungary, among other neighbors. See chapter 4 for discussions relating to the financing and construction of Yugoslavia's Intelsat earth station after 1970.

62. "Tezisy doklada "O mezhdunarodnoi sisteme kosmicheskoi sviazi "Intersputnik" i predstoiaschei konferentsii chlenov 'Intelsat' na zasedanii Soveta "Interkosmos" 20 Dek. 1968," ARAN f. 1678, Sovet "Interkosmos" op. 1, d. 34, ll. 186–87.

63. "Tezisy doklada "O mezhdunarodnoi sisteme kosmicheskoi sviazi "Intersputnik" i predstoiaschei konferentsii chlenov 'Intelsat' na zasedanii Soveta "Interkosmos," ll. 186–187.

64. "Tezisy doklada "O mezhdunarodnoi sisteme kosmicheskoi sviazi "Intersputnik" i predstoiaschei konferentsii chlenov 'Intelsat' na zasedanii Soveta "Interkosmos," ll. 187.

65. Anthony Solomon to Secretary of State, Subject: The INTERSPUTNIK Proposal of the Soviet Union-Action Memorandum, August 30, 1968, 2, Folder 9-1-68, File Economic: TEL 6 7/1/68–11/1/68, 1967–1969 SNF, RG 59, NACP.

66. "Secretary of State to the President, Subject: Soviet INTERSPUTNIK Proposal," September 3, 1968, 1, Folder 9-1-68, Economic: TEL 6 7/1/68–11/1/68, 1967–1969 SNF, RG 59, NACP.

67. Memorandum for Mr Henry A. Kissinger, from Benjamin H. Read, Department of State, "Subject: Forthcoming Conference for Renegotiation of INTELSAT Agreements," February 12, 1969, folder: Communications 1969 and 1970 (2 of 3), Box 313, National Security Files-Subject Files: Communications, Richard Nixon Presidential Library, Yorba Linda, CA.

68. Memorandum for Mr Henry A. Kissinger, 4.

69. "Possible Soviet Intentions re: Participation in the Intelsat Negotiations, and US Responses Thereto," Enclosure to Memorandum for Mr Henry A. Kissinger, from

Benjamin H. Read, Department of State, "Subject: Forthcoming Conference for Rene-gotiation of INTELSAT Agreements," February 12, 1969, Folder: Communications 1969 and 1970 (2 of 3), Box 313, National Security Files–Subject Files: Communica-tions, Richard Nixon Presidential Library, Yorba Linda, CA. These discussions of the possibility of Soviet membership in Intelsat are especially striking for the extent to which they proceeded without consideration for the major Cold War political event of the late summer and fall of 1968: the Soviet invasion of Czechoslovakia. US officials also noted that discussions with the Soviet Union about Intelsat membership would have to be put off temporarily, but expected that they would resume within a few months (Read to Kissinger, 5) Satellite communications thus followed a pattern typi-cal for scientific exchange and economic cooperation during the Cold War, in which high-level geopolitical events largely did not disrupt cooperation and exchange at lower levels. See Autio-Sarasmo, "Stagnation or Not?" 88.

70. Memo from Helmut Sonnenfeldt to Henry Kissinger on "Intelsat Policy," Febru-ary 5, 1970, Folder "Communications 1969+1970, including INTELSAT," Subject Files: Communications, Box 313, NSC Files, Richard Nixon Presidential Library, Yorba Linda, CA.

71. Kildow, *INTELSAT*, 56–57.

72. "Comparison between Interim and Definitive Arrangements, Prepared for Mr. Nicholas Zapple." Eisenhower Presidential Library, Abilene, KS, Abbott Wash-burn Papers, Subseries IV-Intelsat-OTP, Box 109, Folder "Agreements on Definitive Arrangements," 1971 (1), 6.

73. "Comparison between Interim and Definitive Arrangements, Prepared for Mr. Nicholas Zapple," 4.

74. "Memorandum from Charles C. Joyce Jr. to Dr. Kissinger on 'INTELSAT Policy',"
February 4, 1970, NSC Files–Subject Files–Communication, Box 313, Richard Nixon Presidential Library, Yorba Linda, CA.

75. Memorandum from Robert M. Behr to Dr. Kissinger on "INTELSAT Management and International Space Cooperation," January 23, 1970, NSC Files–Subject Files–Communication, Box 313, Richard Nixon Presidential Library, Yorba Linda, CA. In response to these arguments, Charles Joyce and Clay T. "Tom" Whitehead, the Nixon official who helped create the White House's Office for Telecommunications Policy (OTP), again sought to redefine Intelsat in the most politically expedient way, tell-ing their European allies that Intelsat was "a commercial venture and should not be viewed as a prototype for international cooperation in space." "Memorandum from Charles C. Joyce Jr. to Dr. Kissinger on 'INTELSAT Policy'," February 4, 1970, NSC Files–Subject Files–Communication, Box 313, Richard Nixon Presidential Library, Yorba Linda, CA.

76. Memorandum from Robert M. Behr to Dr. Kissinger on "INTELSAT Management and International Space Cooperation."

77. William P. Rogers to President Nixon, March 23, 1971, Space Program Foreign Cooperation 1971, Box 393, Security Council Files, Richard Nixon Presidential Library, Yorba Linda, CA.

78. See chapter 3. On the development of European launch capacity, and especially the French Ariane program and the Guiana Space Center, see Peter Redfield, *Space in the Tropics: From Convicts to Rockets in French Guiana*, (Berkley: University of California Press), 2000.

79. "Summary of Discussion: Post-Apollo Space Cooperation," April 23, 1971, Space Program Foreign Cooperation 1971, Box 393, Security Council Files, Richard Nixon Presidential Library, Yorba Linda, CA, 2.

80. "Technology Transfer in the Post-Apollo Program", July 15, 1971, Space Program Foreign Cooperation (1971) to Staff Secretary, Box 393, Security Council Files, Richard Nixon Presidential Library, Yorba Linda, CA.

81. Henry A. Kissinger to William P. Rogers, July 27, 1971, Space Program Foreign Cooperation (1971) to Staff Secretary, Box 393, Security Council Files, Richard Nixon Presidential Library, Yorba Linda, CA, 2.

82. I. V. Vasil'eva, "Neposredstvennoe sputnikovoe veshchaniie, ego problemy i perspektivy," RGANI F. 5 Tsentral'nyi Komitet KPSS, Otdel transporta i sviazi, op. 63, d. 247, ll. 17–57, 30.

83. Vasil'eva, "Neposredstvennoe sputnikovoe veshchaniie, ego problemy i perspektivy," ll. 31–32.

84. Her various proposals, however, were discussed by a high-level interagency working group organized by the Central Committee and engaging both Sergei Lapin, the head of the State Committee for Television and Radio, and Academician Mstislav Keldysh, head of Interkosmos, among others. See RGANI F. 5, op. 63, d. 247, ll. 58–62.

CHAPTER 4

1. Abbott Washburn, "Status and Outlook of INTELSAT Negotiations," October 6, 1969, Chron October 1969 (2), Subseries IV—INTELSAT-OTP, Box 112, Abbott Washburn Papers, Eisenhower Presidential Library, Abilene, KS.

2. For more on superpower launch capacity as a currency during the Space Race, see chapter 5.

3. A note on terminology: We are using the terms "Earth station" and "ground station" interchangeably throughout this chapter. Other industry terms include "tracking station," dedicated to telemetry data, and "non-geosynchronous satellites." The term "teleport" is sometimes used to point toward the interconnection of the orbital network and terrestrial networks such as internet.

4. James Schwoch, *Global TV: New Media and the Cold War, 1946–69* (Urbana: University of Illinois Press, 2009); James Hay, "The Invention of Air Space, Outer Space,

and Cyberspace," in Lisa Parks and James Schwoch (eds.), *Down to Earth: Satellite Technologies, Industries, and Cultures* (New Brunswick, NJ: Rutgers University Press, 2012), 19–41.

5. Paul N. Edwards, *A Vast Machine: Computer Models, Climate Data, and the Politics of Global Warming* (Cambridge, MA: MIT Press, 2010), 23–25; Paul N. Edwards, "Meteorology as Infrastructural Globalism," *Osiris* 21: 1 (2006), 229–230.

6. Andrew J. Butrica, "Introduction," in Andrew J. Butrica (ed.), *Beyond the Ionosphere: Fifty Years of Satellite Communication* (Washington, DC: NASA, 1997), xxiv. For more on the history of AT&T, its dominant position in American telecommunications, and its fraught relation with the government, see Hugh R. Slotten, *Beyond Sputnik and the Space Race: The Origins of Global Satellite Communications* (Baltimore: Johns Hopkins University Press, 2022).

7. Radio Corporation of America (RCA), "Ground Stations for Space Telecommunications," 1964, 7, folder 8-1-64, file TEL 6, 1964–66 Subject-Numeric Files, RG 59: General Records of the Department of State, National Archives at College Park.

8. RCA, a later entrant into satellite communications technology than AT&T, was greatly encouraged by the US government's decision, based on monopoly concerns, to largely exclude AT&T, despite its significant investment and head start in the technology. For more on AT&T's exclusion, see David Whalen, *The Origins of Satellite Communications, 1945–1965* (Washington, DC: Smithsonian Institution Press, 2002); David J. Whalen, *The Rise and Fall of COMSAT, Technology, Business, and Government in Satellite Communications* (Basingstoke, UK: Palgrave Macmillan, 2014); Slotten, *Beyond Sputnik*.

9. RCA, "Ground Stations," 8.

10. RCA, "Ground Stations," 8–9.

11. For a discussion of secrecy in relation to space science and technology both in the US and Soviet Union, see Andrew Jenks, Securitization and Secrecy in the Late Cold War: The View from Space." *Kritika: Explorations in Russian and Eurasian History* 21: 3 (2020): 659–689.

12. Jenks, "Securitization and Secrecy," 10.

13. COMSAT, "Andover Earth Station," ca. 1967 (Washington, DC: COMSAT Information Office).

14. Abbott Washburn, "Status and Outlook of INTELSAT Negotiations," 7, October 6, 1969, Chron October 1969 (2), Subseries IV—INTELSAT-OTP, Box 112, Abbott Washburn Papers, Eisenhower Presidential Library, Abilene, KS; Jean d'Arcy, "Challenge to Cooperation," *Saturday Review*, October 24, 1970, 25. Articles (4), Subseries IV—INTELSAT-OTP, Box 110, Abbott Washburn Papers, Eisenhower Presidential Library, Abilene, KS; "The World's Earth Stations for Satellite Communications," *COMSAT*, December 1970, 1, Comsat (2), Subseries IV—INTELSAT-OTP, Box 116, Abbott Washburn Papers, Eisenhower Presidential Library, Abilene, KS.

15. Of course, with the introduction of direct broadcast satellites, this changed significantly since everyone with a properly oriented satellite dish would be able to pick up the signal.

16. William G. Geddes (Director, Operations, INTELSAT) "The Live Via Satellite Era," in *Intelsat Memoirs* (Washington, DC: International Telecommunications Satellite Organization, 1979). See also chapter 1 of this book.

17. The map projection is thus similar to the networks maps used for describing the technical links used during the "Our World" broadcast, as described in chapter 2.

18. COMSAT, "The Global Communications Satellite System," February 1971, Washburn Papers, Subseries IV, INTELSAT-OTP, Box 116, Eisenhower Library, Abilene, KS.

19. P.T. Indonesian Satellite Corporation "Indonesian Earth Satellite Station," 1969, folder 10/1/69, TEL 6, Space Communications, 1967–1969 SNF, RG59, NACP.

20. "Vladivostok-Kosmos-Moskva," *Tekhnika Molodezhi*, No. 7 (July 1965): 14.

21. Illustration by A. Minenkov, *Aviatsiia i Kosmonavtika* No. 6 (June 1969), rear cover.

22. Charles R. Denny, Vice President, RCA, to U. Alexis Johnson, Deputy Undersecretary of State, June 7, 1966, File TEL 6, 1964–1969 SNF, RG59, NACP.

23. The prioritized countries were Colombia, Chile, Brazil, Nigeria, Ethiopia, Kenya, Tanzania, Uganda, Turkey, Pakistan, India, Thailand, Philippines, South Korea, and one Earth station serving all six countries in Central America. Benjamin Read, Executive Secretary, Department of State, to Walter Rostow, White House, "Memorandum on Communication Satellite Earth Station Construction in Less Developed Countries, NSAM 342," December 31, 1966, Folder 1/1/67, File TEL 6 Space Communications, 1967–1969 SNF, RG59, NACP.

24. Read, "Memorandum on Communication Satellite Earth Station Construction in Less Developed Countries," 2.

25. Read, "Memorandum on Communication Satellite Earth Station Construction in Less Developed Countries."

26. Institute of Electrical and Electronics Engineers (IEEE), "Conference Publication No. 72," *Earth Station Technology, Proceedings of the IEEE Conference on Earth Station Technology*, October 14–16, 1970, Savoy Place, London.

27. See, for example, Leland Johnson, "The Commercial Uses of Communication Satellites," *California Management Review*, 5: 3 (1963), 55–66.

28. Note again the colonizing erasure of the highly salient and specific geography and history of countries of the Global South constructing Earth stations in this period. L. B. Early, L. Kumins, and J. Baer, "Business Forecasting for Communication Satellite Systems," in Richard B. Marsten (ed.), *Communication Satellite Systems Technology* (New York: Academic Press, 1966), 941–954.

29. President John F. Kennedy to Lyndon B. Johnson, President of the Senate, and John W. McCormack, Speaker of the House of Representatives, February 6, 1962. JFK Speech Files, Theodore C. Sorenson, Personal Files. Box 67. JFK Library. On "footprint analysis," see Lisa Parks, "Satellites, Oil, and Footprints: Eutelsat, Kazat, and Post-Communist Territories in Central Asia," in Lisa Parks and James Schwoch (eds.), *Down to Earth: Satellite Technologies, Industries, and Cultures* (New Brunswick, NJ: Rutgers University Press, 2012), 122–137; see also Ingrid Volkmer, "Satellite Cultures in Europe: Between National Spheres and a Globalized Space," *Global Media and Communication* 4: 3 (2008), 231–244.

30. The idea of a bilateral, "country by country" expansion of Earth station infrastructure, however, has come down to us in the historiography of satellite communications. See, for example, Butrica, "Introduction," xvi.

31. As Whalen notes, one of the major reasons that NASA intervened in the private sector's development of communications satellites was to be able to claim that investment in space technology could be profitable (and thus help justify and offset the enormous expenditures on human spaceflight in the 1960s). Whalen, *The Origins of Satellite Communications*, 18.

32. US Department of Commerce, Office of Telecommunications, "An Analysis of Domestic and Foreign Small Earth Station Markets," OT Contractor Report 76–3, May 1976.

33. US Embassy Belgrade to Department of State, Airgram A-84, August 3, 1967, Tel 6, 1967–1969 SNF, RG 59, NACP. The article referred to in the airgram does not state which system Yugoslavia was going to join, but it does note the countries' favorable geopolitical position, and the preamble reads: "Satellite station for four countries—Bulgaria, Romania, Hungary and Yugoslavia to have a joint center—According to proposal, installations would be mounted in Yugoslavia," "New Relations of Yugoslav PTT Enterprises", *Politika*, Friday, July 14, 1967, 8.

34. US Embassy Belgrade to Secretary of State, Telegram 2376, March 7, 1968, Tel 6, 1967–1969 SNF, RG 59, NACP.

35. US Embassy Belgrade to Secretary of State, "Intelsat," Telegram 2584, March 27, 1968, Tel 6, 1967–1969 SNF, RG 59, NACP.

36. US Embassy Belgrade to Secretary of State, "Intelsat."

37. US Embassy Belgrade to Secretary of State, "Intelsat."

38. "Annex I: Earth Station Development," *Communications Satellite Earth Station Construction in Less Developed Countries—NSAM 342*, December 31, 1966, Tel 6, 1967–1969 SNF, RG 59, NACP.

39. Intersputnik, *Intersputnik: Chronicling a Long and Glorious Path* (Moscow: Smit and Hartman, 2011).

40. Lukasz Stanek, *Architecture in Global Socialism: Eastern Europe, West Africa, and the Middle East in the Cold War* (Princeton, NJ: Princeton University Press, 2020).

41. Interview with František Šebek, Prague, May 25, 2018.

42. K stoletiiu so dnia rozhdeniia Alekseia Fedorovicha Bogomolova. Kniga 2. Ocherki razvitiiia OKB MEI v litsakh. Period 1965–1988 [On the 100th anniversary of the birth of Aleksei Fedorovich Bogomolov. Book 2. Sketches of the development of OKB MEI in its people from the period 1965–1988], Moscow: AO OKB MEI, 2015, 69.

43. RGAE F. 3527, op. 72, d. 579, "Dokumenty ob uchastii MinSviazi v rabote MOKS Intersputnika," l. 1. In addition to suggesting that the origins of space industry privatization lie long before 1991, this picture of enthusiastic but still relatively secretive (hence payment was concealed by being routed through Bermudan banks) Soviet commercial space activity offers an interesting counterpart to Andrew Jenks's findings that, by the 1980s, Soviet internationalism in its civilian space activities stood in contrast to an increasingly paranoid and isolationist US space program. Jenks, "Securitization and Secrecy". The story of active and influential Soviet involvement in global satellite communications is largely missing from the literature. See, for example, Schwoch, *Global TV*, 153; Similarly, Diana Lemberg, in her otherwise excellent history of the US pursuit of the "free flow of information," proposes that the Soviet Union, "[h]aving lost its early lead in the space race, appeared somewhat less interested in satellite communications than the United States, probably because the technology was less essential for communications in its contiguous Eurasian sphere of influence." Diana Lemberg, *Barriers Down: How American Power and Free-Flow Policies Shaped Global Media* (New York: Columbia University Press, 2019), 145.

44. Jose Altschuler, "From Short Wave and Scatter to Satellite: Cuba's International Communications," in Andrew Butrica (ed.), *Beyond the Ionosphere: Fifty Years of Satellite Communication* (Washington, DC: NASA, 1997), 248.

45. Intersputnik, *Intersputnik: Chronicling a Long and Glorious Path*; "Soviet Intelsat Accord Enters Force," *Sel'skaia zhizn'* October 4, 1985, 3, in FBIS No. 203, NASA Archives, Record No. 15738, Series: International Cooperation and Foreign Countries, Subseries "International Cooperation," Folder "Intelsat."

46. On earlier satellite-tracking stations in Kenya, Madagascar, and elsewhere, see Asif Siddiqi, "Shaping the World: Soviet-African Technologies from the Sahel to the Cosmos," *Comparative Studies of South Asia, Africa and the Middle East* 41: 1 (2021), 41–55; and Lisa Parks, "Global Networking and the Contrapuntal Node: The Project Mercury Earth Station in Zanzibar, 1959–64," *Zeitschrift für Medien- und Kulturforschung* 11: 1 (2020), 40–57.

47. Theodore L. Elliot to Henry A. Kissinger, "Proposed Telephone Conversation between President Nixon and King Hassan of Morocco; Enclosure: Talking Points," January 5, 1970. Box 1563, Tel 6, Economic, 1970–1973 SNF, RG 59, NACP. For more on this moment, see Slotten, *Beyond Sputnik*, 183–184.

48. Letter from J. D. O'Connell to Anthony Solomon, August 23, 1968, Tel 6 Space Communications, 1967–1969 SNF, RG59, NACP, 1–2.

49. Letter from Abbott Washburn to Leonard H. Marks, November 18, 1969. Washburn Papers; Subseries IV, Intelsat—OTP; Box 112, Eisenhower Presidential Library, Abilene, KS.

50. "Communication Satellites: Technical, Economic, and International Developments." Staff Report Prepared for the Use of the Committee on Aeronautical and Space Sciences, United States Senate. February 25, 1962. US Government Printing Office, Washington, DC, 1962, 118 (emphasis in the original).

51. Letter from Washburn to Marks.

52. Staffan Ericson and Kristina Riegert. *Media Houses: Architecture, Media and the Production of Centrality* (New York: Peter Lang, 2010); Tung-Hui Hu. *A Prehistory of the Cloud* (Cambridge, MA: MIT Press, 2015); Bruce R. Elbert, *The Satellite Communication Ground Segment and Earth Station Handbook* (Boston: Artech House, 2001), 323f.

53. The fact of widespread representations of satellite Earth stations when they were a new technology highlights Brian Larkin's argument that, rather than only being "visible on breakdown," following Susan Leigh Star and others, in fact infrastructures can be invisible or spectacular, or anywhere between those two poles. Susan Leigh Star, "The Ethnography of Infrastructure," *American Behavioral Scientist* 43: 3 (1999), 380, 382; Brian Larkin, "The Politics and Poetics of Infrastructure," *Annual Review of Anthropology* 42: 1 (2013), 336.

54. All stamps reproduced and analyzed in this discussion are from the authors' personal collection. Donald M. Reid points toward postage stamps as a source for historians, as physical objects, as evidence of the existence of a postal service, and also as bearers of symbols. Reid also notes that few people can describe the stamp on a letter that they received the day before, that stamps, just as other media infrastructures, are "all but invisible." Donald M. Reid, "The Symbolism of Postage Stamps: A Source for the Historian," *Journal of Contemporary History* 19: 2 (1984), 223.

55. The examples discussed here are typically so-called commemorative (as opposed to definitive) stamps, depicting historical landmarks "with strong political iconic messages." Jack Child, *Miniature Messages: The Semiotics and Politics of Latin American Postage Stamps* (Durham, NC: Duke University Press, 2008), 17.

56. Svetlana Boym, "Kosmos: Remembrances of the Future," in Adam Bartos and Svetlana Boym (eds.), *Kosmos: A Portrait of the Russian Space Age* (Princeton, NJ: Princeton University Press, 2001), 83.

57. See, for example, Igor Cusack, "Tiny Transmitters of Nationalist and Colonial Ideology: The Postage Stamps of Portugal and Its Empire," *Nations and Nationalism* 11: 4 (2005), 591–612; Pauliina Raento, "Introducing Popular Icons of Political Identity," *Geographical Review* 101: 1 (2011), iii–vi; Christopher B. Yardley, *The Representation of Science and Scientists on Postage Stamps: A Science Communication Study* (Canberra: Australian National University Press, 2015); Child, *Miniature Messages*.

58. COMSAT, "The World's Earth Stations for Satellite Communications," December 1970. Abbott Washburn Papers, Subseries IV-Intelsat-OTP, Box 116, Folder COMSAT (2), Eisenhower Presidential Library, Abilene, KS.

59. Urmila Devgon, "Intelsat: A Cooperative Partnership in Communications," *Topic: Special Issue: Life in the 21st Century* 52, 1970, 25.

60. See, for example, "Protokol VI Sessii g. Varshava, 1–8 oktiabria 1977 g," Intersputnik Organization Corporate Archives, Moscow, 10.

61. Intersputnik, *Intersputnik: Chronicling a Long and Glorious Path.*

62. For more information on Eurasian postal networks, see Marsha Siefert, "'Chingis-Khan with the Telegraph': Communication in the Russian and Ottoman Empires," in Jörn Leonhard and Ulrike Hirschhausen (eds.) *Comparing Empires: Encounters and Transfers in the Long Nineteenth Century* (Göttingen: Vandenhoeck & Ruprecht, 2012).

CHAPTER 5

1. Interview with František Šebek in Prague, May 25, 2018. Our sincere thanks to Dr. Šebek for his time, his hospitality, and for an unforgettable evening in Prague.

2. See Bohdan Shumylovych (text) and Olha Povoroznyk (video), "Future from the Past: Imaginations on the Margins," https://ars.electronica.art/keplersgardens/de /imaginations/, September 12, 2020 (accessed March 27, 2023).

3. Sari Autio-Sarasmo, "Stagnation or Not? The Brezhnev Leadership and the East-West Interaction," in Dina Fainberg (ed.), *Reconsidering Stagnation in the Brezhnev Era: Ideology and Exchange* (Lanham, MD: Lexington Books, 2016); Sari Autio-Sarasmo, "Technological Modernisation in the Soviet Union and Post-Soviet Russia: Practices and Continuities," *Europe-Asia Studies* 68: 1 (2016), 79–96.

4. Per Högselius, *Red Gas: Russia and the Origins of European Energy Dependence* (New York: Palgrave Macmillan, 2013).

5. Audra J. Wolfe, *Freedom's Laboratory: The Cold War Struggle for the Soul of Science* (Baltimore: Johns Hopkins University Press, 2018).

6. Andrew Jenks has made a similar argument about international cooperation in human spaceflight after Apollo–Soyuz. Andrew Jenks, "Securitization and Secrecy in the Late Cold War: The View from Space." *Kritika: Explorations in Russian and Eurasian History* 21: 3 (2020): 659–689.

7. Jenks, "Securitization and Secrecy"; Andrew Jenks, "U.S.-Soviet Handshakes in Space and the Cold War Imaginary," *Journal of Cold War Studies* 23: 2 (2021): 100–132; Thomas Ellis, "Howdy Partner!" Space Brotherhood, Detente and the Symbolism of the 1975 Apollo-Soyuz Test Project," *Journal of American Studies* 53: 3 (2019), 744–769; Darina Volf, "Evolution of the Apollo-Soyuz Test Project: The Effects of the

'Third' on the Interplay Between Cooperation and Competition," *Minerva* 59 (2021): 399–418; Debbora Battaglia, "Arresting Hospitality: The Case of the 'Handshake in Space'," *Journal of the Royal Anthropological Institute* 18: 1 (2021), 76–89.

8. David Reynolds, *Summits: Six Meetings That Shaped the 20th Century* (New York: Basic Books, 2009), 235.

9. On "space brotherhood," see Ellis, "Howdy Partner!."

10. *Department of State Newsletter*, no. 198, February 1978, 11, https://books .google.com.ua/books?id=TARvyxfVAC0C&pg=PA1&hl=uk&source=gbs_selected _pages&cad=3#v=onepage&q&f=false (accessed July 29, 2021).

11. Richard A. Moss, *Nixon's Back Channel to Moscow: Confidential Diplomacy and Détente* (Lexington: University Press of Kentucky, 2017).

12. "Meeting between Presidential Assistant Kissinger and Ambassador Dobrynin. Document 162 Memorandum of Conversation (USSR)" Washington, DC, May 28, 1971, in *Soviet-American Relations: The Détente Years, 1969–72*, eds. David C. Geyer and Douglas E. Selvage, Department of State Publication 11438 (Washington, DC: US Government Printing Office, 2007), 368.

13. The fact that the Soviet Union obtained Intelsat Earth stations via these hotline negotiations meant that their integration into Intelsat's network (and the satellite communications private sector) looked quite different from China's. China also gained an Earth station during this era of high-level summitry to provide television coverage for President Nixon's visit to Beijing in 1972. However the Earth station needed for this coverage was a temporary one, delivered and installed by the US government. Thomas O'Toole, "Chinese May Plug into Satellite System," *Washington Post*, January 27, 1972, A1.

14. *Department of State Newsletter*, 11; Delbert D. Smith, *Communication via Satellite* (Leiden: A. W. Sijthoff, 1976), 288, n. 101.

15. "Essential Elements of the US-USSR Direct Communications Link Implementation Plan," November 30, 1971, Communications 1971, National Security Council Files, Box 313, Richard Nixon Presidential Library, Yorba Linda, CA; Desmond Ball, "Improving Communications Links between Moscow and Washington," *Journal of Peace Research* 38: 2 (1991), 140.

16. Ball, "Improving Communications Links." Ball also notes that, beyond the redundancy of the cable and microwave circuits on the ground in each capital city, the Intelsat and Gorizont satellites themselves, as well as their Earth stations, were entirely vulnerable to destruction in a nuclear war.

17. Ball, "Improving Communications Links," 140.

18. Ball, "Improving Communications Links," 141.

19. "US-Soviet Communications: An Eyewitness Report," Office of Telecommunications Policy (OTP), Subseries IV, Intelsat, Box 120, Washburn Papers, Dwight D.

Eisenhower Library, Abilene, KS; "I.T.T. Gets Hotline Contract," *New York Times*, November 7, 1972, 55.

20. "US-Soviet Communications: An Eyewitness Report."

21. See, for example, a report on a series of meetings with the Japanese firm NES, which worked extensively for INTELSAT and COMSAT, in June and July 1972, followed by the decision to purchase satellite communications components from them. RGAE f. 3527, op. 63, d. 1660 "Materialy po mezhdunar. Sotrudnichestvu," 1972, ll. 97–106. These contacts were not an entirely new development—the graduate expansion of Soviet telecommunications networks to both the US and the rest of the world had been taking place since the cable DCL in 1963. "US-Soviet Communications: An Eyewitness Report."

22. Jenks, "Securitization and Secrecy"; Oscar Sanchez-Sibony, *Red Globalization: The Political Economy of the Soviet Cold War from Stalin to Khrushchev*, (Cambridge: Cambridge University Press, 2014).

23. "Memorandum for the Honorable Henry A. Kissinger, Assistant to the President for National Security Affairs, The White House, Subject: 'US-USSR Direct Communications Link'," November 30, 1971, Communications 1971, National Security Council Files, Box 313, Richard Nixon Presidential Library, Yorba Linda, CA.

24. "Essential Elements of the US-USSR Direct Communications Link Implementation Plan," November 30, 1971, Communications 1971, National Security Council Files, Box 313, Richard Nixon Presidential Library, Yorba Linda, CA, 4.

25. "Essential Elements of the US-USSR Direct Communications Link Implementation Plan," 2.

26. "Defense Communications Agency Implementation Plan for Improved US–USSR Direct Communications Link," Annex C, Discussion of Policy and Legal Issues, November 29, 1971, Communications 1971, National Security Council Files, Box 313, Richard Nixon Presidential Library, Yorba Linda, CA.

27. An exhibit on the history of satellite communications made this point in a panel about the satellite DCL. "Sputnikovaia sviaz': v kosmose i na zemle" exhibit, Museum of Cosmonautics, Moscow, November 22, 2018–November 25, 2019.

28. "Essential Elements of the US-USSR Direct Communications Link Implementation Plan," 3.

29. RGAE f. 3527 op. 63 delo 780, l. 54.

30. Stephen Graham and Nigel Thrift, "Out of Order: Understanding Repair and Maintenance," *Theory, Culture & Society* 4: 3 (2007), 1–25; Christopher R. Henke and Benjamin Sims, *Repairing Infrastructures: The Maintenance of Materiality and Power* (Cambridge, MA: MIT Press, 2020). For an example of maintenance work, see Julia Velkova on security, labor, and maintenance at a Yandex data center. Julia Velkova, "The Art of Guarding the Russian Cloud: Infrastructural Labour in a Yandex Data

Centre in Finland," *Studies in Russian, Eurasian and Central European New Media* 20 (2020), 47–63.

31. For example, an Earth Station Technology seminar, organized by COMSAT, was held in June 1976 in Munich.

32. See, for example, "United Kingdom Seminar on Communications Satellite Earth Station Planning and Operation," London 1968, 10–17; IEEE Conference on Earth Station Technology, October 14–16, London, 1970, 374–380.

33. Asif Siddiqi, "Dispersed Sites: San Marco and the Launch from Kenya," in John Krige (ed.), *How Knowledge Moves: Writing the Transnational History of Science and Technology* (Chicago: University of Chicago Press, 2019), 175–200; Lisa Parks, "Global Networking and the Contrapuntal Node: The Project Mercury Earth Station in Zanzibar, 1959–64," *Zeitschrift für Medien- und Kulturforschung* 11: 1 (2020), 40–57.

34. "United Kingdom Seminar on Communications Satellite Earth Station Planning and Operation," 1968, Section F, Paper 3, fig. 5.

35. J. B. Potts, "Introduction," in *Proceedings of INTELSAT V Earth Station Technology Seminar*, June 13–16, 1976, Munich, Federal Republic of Germany, I-1.

36. Potts, "Introduction"; see also "Introduction," *INTELSAT Second Earth Station Seminar*, Athens, October 24–26, 1977.

37. *Proceedings of INTELSAT V Earth Station Technology Seminar*, June 13–16, 1976, Munich, Federal Republic of Germany, I-2.

38. See, for example, "The INTELSAT Decade, 1965–1974," 20–21, Abbott Washburn Papers, Subseries IV: INTELSAT-OTP, Box 110, Eisenhower Presidential Library, Abilene, KS. This commemorative brochure lists the non-Intelsat Soviet Earth station outside Moscow as nonstandard, along with French, Saudi, and Egyptian Earth stations.

39. See chapter 4 of this book. On US resistance to distributing Soviet television and film within the US, see Tony Shaw and Denise J. Youngblood, *Cinematic Cold War: The American and Soviet Struggle for Hearts and Minds* (Lawrence: University Press of Kansas, 2014).

40. RGAE F. 3527, op. 63 d. 3414 "Materialy po mezhdunarodnomu sotrudnich-estvu," l. 8.

41. RGAE F. 3527, op. 63, d. 3418, "To zhe na razrabotku tekhnicheskikh reshenii po dooborudovaniiu zemnoi stantsii sputnikovoi sviazi v raione gor. L'vova," December 2, 1974, ll. 1–3.

42. Ellen Propper Mickiewicz, *Split Signals: Television and Politics in the Soviet Union* (Oxford: Oxford University Press, 1988), 68–79. For more on Soviet and American journalists in the 1970s and 1980s, see Dina Fainberg, *Cold War Correspondents: Soviet and American Reporters on the Ideological Frontlines* (Baltimore: Johns Hopkins University Press, 2021).

43. Russian State Archive of Economics (RGAE) F. 3527 Ministerstvo Sviazi, op. 72, d. 665 "Ekspertnoe zakliuchenie No. 3358 po rabochemu proektu na stroitel'stvo kosmicheskoi stantsii sistemy 'Intersputnik' v Nikaragua," March 16, 1984 ll. 1–3.

44. Costs for telephone and television transmissions via Intersputnik were significantly lower than via Intelsat. John Downing, "International Communications and the Second World: Developments in Communication Strategies," *European Journal of Communication* 4: 1 (1989), 115–116.

45. Russian State Archive of Economics (RGAE) F. 3527 Ministerstvo Sviazi, op. 72, d. 665 "Ekspertnoe zakliuchenie No. 3358 po rabochemu proektu na stroitel'stvo kosmicheskoi stantsii sistemy 'Intersputnik' v Nikaragua," March 16, 1984 ll. 1–3.

46. RGAE F. 3527 Ministerstvo Sviazi, op. 72, d. 665 "Ekspertnoe zakliuchenie No. 3358," l. 4.

47. Lisa Parks and Nicole Starosielski, "Introduction," in Lisa Parks and Nicole Starosielski (eds.), *Signal Traffic: Critical Studies of Media Infrastructure* (Urbana: University of Illinois Press, 2015), 2–3. James Schwoch makes a similar argument with regard to the US telegraph system; see James Schwoch, *Wired into Nature. The Telegraph and the North American Frontier* (Urbana: University of Illinois Press, 2018).

48. "Soviet Intelsat Accord Enters Force," *Sel'skaia zhizn'* October 4, 1985, 3, in FBIS No. 203, NASA Archives, Record No. 15738, Series: International Cooperation and Foreign Countries, Subseries "International Cooperation," Folder "Intelsat."

49. G. N. Pashkov, "O tendentsiiakh razvitiia tekhniki dlia dal'neishego osvoeniia kosmicheskogo prostranstva v Sodeninennykh statakh ameriki," RGANI F. 5, op. 66, d. 305, l. 73. See Chapter 3. "The Soviet Statsionar Satellite Communications System: Implications for INTELSAT," Interagency Intelligence Memorandum, NIO IMM 76–016, April 1976, CIA: 1. https://www.cia.gov/readingroom/docs/DOC _0000283805.pdf (accessed April 30, 2022).

50. RGAE f. 3527, op. 63, d. 8806, "Protokol 10 sessii Soveta Mezhdunarodnoi organizatsiia kosmich sviazi "Intersputnik" g. Brno October 1981," l. 7.

51. RGAE f. 3527, op. 63, d. 8806, "Protokol 10 sessii," l. 10.

52. "RGAE F. 3527, op. 72, d. 579, "Dokumenty ob uchastii MinSviazi v rabote MOKS Intersputnika," l. 1.

53. "RGAE F. 3527, op. 72, d. 579, "Dokumenty ob uchastii," l. 1.

54. Elizabeth Tucker, "CNN, Soviets Negotiating TV Program Pact," *Washington Post*, August 22, 1984, A-1, A-21. RGAE F. 3527, op. 72, d. 3065, "Protokol soveshchaniia ekpertov . . . Intersputnik po ekspluatatsii mezh. Sistemy sputnikovoi sviazi "Intersputnik," Zruch nad Sazavou, March 1988," l. 23.

55. RGAE F. 3527, op. 72, d. 3065, "Protokol soveshchaniia ekpertov . . . Intersputnik po ekspluatatsii mezh. Sistemy sputnikovoi sviazi 'Intersputnik,' Zruch nad Sazavou, March 1988," l. 32.

EPILOGUE

1. Abbott Washburn, "Separate International Satellite Systems Raise Profound U.S. Policy Question," *Telematics and Informatics* 1: 4 (1984), 448 (emphasis in original).

2. Washburn, "Separate International Satellite Systems," 450.

3. Joseph N. Pelton, "Intelsat and the Article XIV (d) Test of Significant Economic Harm or, If Someone Is Dead Would You Ask If They Were Significantly Wounded?," presentation at the Satellite Communications Users Conference, August 22–24, 1983 (St. Louis, MO, 1983), 18.

4. Pelton, "Intelsat and the Article XIV(d)", 14.

5. "Statement of Irving Goldstein, President, Communications Satellite Corporation, before the House Subcommittee on Telecommunications and Finance, Committee on Energy and Commerce, June 13, 1984," Abbott Washburn Papers, Box 217, Folder: Separate Systems 1984 (3), Dwight D. Eisenhower Presidential Library, Abilene, KS.

6. "Statement of Irving Goldstein," 3.

7. "Statement of Irving Goldstein," 2.

8. "Statement of Irving Goldstein," 3.

9. "Statement of Irving Goldstein."

10. Abbott Washburn, "Statement of Abbott Washburn before a Joint Meeting of the Subcommittee on International Operations and the Subcommittee on International Economic Policy and Trade, Committee of Foreign Affairs, US House of Representatives, Washington D.C.," March 6, 1985; Washburn Papers; Subseries VII. Intelsat Consultant. 1982–88; Box 221; Eisenhower Library, Abilene, KS, 18.

11. See, for example, Ellen Propper Mickiewicz, *Split Signals: Television and Politics in the Soviet Union* (Oxford: Oxford University Press, 1988), 13–16.

12. Robert Campbell, "Satellite Communications in the USSR," *Soviet Economy* 4: 1 (1985), 336.

13. These developments paralleled the Soviet decision to expand international cooperation after the Apollo–Soyuz mission, just as the US retreated from international cooperation in space, as Andrew Jenks has argued. Andrew L. Jenks, *Collaboration in Space and the Search for Peace on Earth*, Anthem Series on Russian, East European, and Eurasian Studies (New York: Anthem Press, 2021).

14. Robina Riccitiello, "Intersputnik Links East to West: Organization Builds Its Customer Base by Offering Low Prices," *Space News*, October 5–11, 1992.

15. "The Geopolitics of 5G: America's War on Huawei Nears Its Endgame," *The Economist*, July 16, 2020. https://www.economist.com/briefing/2020/07/16/americas-war-on-huawei-nears-its-endgame (accessed March 28, 2022). The US was preceded by Australia and Japan, which had put a ban on Huawei already in 2018.

16. In 2020, the five largest telecommunications manufacturing companies were Huawei (China), Nokia (Finland), Ericsson (Sweden), ZEN (China), and Cisco (US). "The Geopolitics of 5G," *The Economist*, https://www.economist.com/briefing/2020 /07/16/americas-war-on-huawei-nears-its-endgame (accessed March 28, 2022).

17. Milton L. Mueller, *Networks and States: The Global Politics of Internet Governance* (Cambridge, MA: MIT Press, 2010), 3.

18. Initiatives to provide service to remote areas around the planet has not been limited to satellite access. Google's Project Loon, for instance, used a large-scale network of balloons floating in the stratosphere in order to distribute signal flows to internet users disconnected from terrestrial infrastructures. For a discussion on Project Loon as an example of elemental infrastructures, see Derek P. McCormack, "Elemental Infrastructures for Atmospheric Media: On Stratospheric Variations, Value and the Commons," *Environment and Planning D: Society and Space* 35: 3 (2017), 418–437; Hannah Zindel, "Ballooning: Aeronautical Techniques from Montgolfier to Google," in Jörg Dünne, Kathrin Fehringer, Kristina Kuhn, and Wolfgang Struck (eds.), *Cultural Techniques: Assembling Spaces, Texts and Collectives* (Berlin: De Gruyter, 2020), 107–127.

19. https://www.starlink.com/resources, accessed March 31, 2023.

20. Mike Wall, "China's Tianhe Space Station Module Dodged SpaceX Starlink Satellites Twice This Year," Space.com, December 30, 2021, https://www.space.com/china -tianhe-space-station-maneuvers-spacex-starlink. In February 2022, the National Aeronautics and Space Administration (NASA) raised similar concerns over the next generation of Starlink low-Earth-orbit satellites, Elisabeth Howell, "NASA Is Concerned about SpaceX's New Generation of Starlink Satellites," Space.com, February 10, 2022, https://www.space.com/nasa-collision-risk-starlink. For a discussion of space debris and the reshaping of orbital and planetary environments with the entry of commercial actors, and how eventually infrastructures will put other infrastructures at risk. See Michael Clormann and Nina Klimburg-Witjes, "Troubled Orbits and Earthly Concerns: Space Debris as a Boundary Infrastructure," *Science, Technology, & Human Values* 47: 5 (2022), 960–985.

21. Daniel Clery, "Starlink Already Threatens Optical Astronomy. Now, Radio Astronomers Are Worried," *Science*, October 9, 2020. https://www.science.org/content/article /starlink-already-threatens-optical-astronomy-now-radio-astronomers-are-worried

22. On the very first day of Russia's war against Ukraine, on February 24, 2022, the satellite internet provided by Viasat (using a single KA-SAT satellite rather than a microsatellite system) was disrupted by a cyberattack, "against the satellite's ground infrastructure—not the satellite itself," and users across Europe lost their connection, among them Ukraine's defenses. Matt Burgess, "A Mysterious Satellite Hack Has Victims Far beyond Ukraine," *Wired*, March 23, 2022. https://www.wired.com /story/viasat-internet-hack-ukraine-russia/

23. Initial claims by SpaceX suggested that it provided Ukraine with Starlink terminals without compensation. It was later reported, however, that the US government,

via the US Agency for International Development, paid for roughly a third of the 5,000 terminals. See Cristiano Lima, "U.S. Quietly Paying Millions to Send Starlink Terminals to Ukraine, Contrary to SpaceX Claims," *Washington Post*, April 8, 2022. https://www.washingtonpost.com/politics/2022/04/08/us-quietly-paying-millions -send-starlink-terminals-ukraine-contrary-spacexs-claims/. See also Alex Marquardt, "Musk's SpaceX Says It Can No Longer Pay for Critical Satellite Services in Ukraine, Asks Pentagon to Pick up the Tab," CNN, October 14, 2022. https://www.cnn.com /2022/10/13/politics/elon-musk-spacex-starlink-ukraine

24. Rachel Lerman and Cat Zakrzewski, "Elon Musk's Starlink Is Keeping Ukrainians Online When Traditional Internet Fails," *Washington Post*, March 19, 2022. https://www.washingtonpost.com/technology/2022/03/19/elon-musk-ukraine-starlink/

25. Starlink website (https://www.starlink.com/satellites), March 29, 2022.

26. SWP is one of Europe's largest think tanks and advises the German government, as well as decision-makers in the rest of the European Union, the North Atlantic Treaty Organization (NATO), and the United Nations.

27. Daniel Voelsen, *Internet from Space: How New Satellite Connections Could Affect Global Internet Governance,* (SWP Research Paper, 3/2021) (Berlin: Stiftung Wissenschaft und Politik—SWP—Deutsches Institut für Internationale Politikund Sicherheit, 2021).

28. Voelsen, *Internet from Space*, 6.

29. Voelsen, *Internet from Space*, 6.

30. Susan Leigh Star and Karen Ruhelder, "Steps toward an Ecology of Infrastructure: Design and Access for Large Information Spaces," *Information Systems Research* 7: 1 (1996), 114.

BIBLIOGRAPHY

ARCHIVES

Arkhiv Rossiiskoi Akademii Nauk (ARAN) [Archive of the Russian Academy of Sciences], Moscow, Russia.

BBC Written Archives Centre, Caversham, UK.

British Telecom Archives, London.

Dwight D. Eisenhower Presidential Library Archives, Abilene, KS.

European Broadcasting Union Archives, Geneva, Switzerland.

Gosudarstvennyi Arkhiv Rossiiskoi Federatsii (GARF) [State Archive of the Russian Federation], Moscow.

Intersputnik Archives, Moscow.

John F. Kennedy Presidential Library Archives, Boston.

NASA Headquarters Archives, Washington, DC.

National Archive of Contemporary History, Tbilisi, Georgia.

National Archives and Records Administration (NARA), College Park, MD.

Richard M. Nixon Presidential Library Archives, Yorba Linda, CA.

Rossiiskoi Gosudarstvennyi Arkhiv Ekonomiki (RGAE) [Russian State Archive of Economics], Moscow.

Rossiiskoi Gosudarstvennyi Arkhiv Noveishei Istorii (RGANI) [Russian State Archive of Contemporary History], Moscow.

PUBLISHED WORKS

Adzhubei, Aleksei, and L. P. Grachev (eds). 1960. *Odin den' mira: Sobytiia 27 sentiabria 1960 g.* Moscow: Izvestiia.

Alonso, Pedro Ignacio. 2016. "Introduction: Towards an Archaeology of Things Moving." In *Space Race Archaeologies: Photographs, Biographies, and Design*, edited by Pedro Ignacio Alonso, 7–20. Berlin: DOM Publishers.

Alonso, Pedro Ignacio, and Hugo Palmarola. 2017. "NASA in Chile: Technology and Branding of a Satellite-Tracking Station." *Design Issues* 33 (2): 31–42.

Altschuler, Jose. 1997. "From Short Wave and Scatter to Satellite: Cuba's International Communications." In *Beyond the Ionosphere: Fifty Years of Satellite Communication*, edited by Andrew Butrica, 243–249. Washington, DC: National Aeronautics and Space Administration (NASA).

Anand, Nikhil, Akhil Gupta, and Hannah Appel (eds.). 2018. *The Promise of Infrastructure*. Durham, NC: Duke University Press.

Andrews, James T., and Asif Siddiqi. 2011. *Into the Cosmos: Space Exploration and Soviet Culture*. Pittsburgh: Pittsburgh University Press.

Arendt, Hannah. 1958. *The Human Condition*. 2nd ed. Chicago: University of Chicago Press.

Arendt, Hannah. 2007. "The Conquest of Space and the Stature of Man." *The New Atlantis: A Journal of Technology and Society*, no. 18, Fall 2007, 43–55.

Auslander, Philip. 2008. *Liveness: Performance in a Mediatized Culture*. London: Routledge.

Autio-Sarasmo, Sari. 2016. "Stagnation or Not? The Brezhnev Leadership and the East-West Interaction." In *Reconsidering Stagnation in the Brezhnev Era: Ideology and Exchange*, edited by Dina Fainberg and Artemy Kalinovsky, 87–103. Lanham, MD: Lexington Books.

Autio-Sarasmo, Sari. 2016. "Technological Modernisation in the Soviet Union and Post-Soviet Russia: Practices and Continuities." *Europe-Asia Studies* 68 (1): 79–96.

Autio-Sarasmo, Sari. 2018. "Planning in Cold War Europe: Competition, Cooperation, Circulations (1950s–1970s)." In *Planning in Cold War Europe: Competition, Cooperation, Circulations (1950s–1970s)*, edited by Michael Christian, Sandrine Kott, and Ondrej Matejka, 143–164. Berlin: De Gruyter Oldenbourg.

Badenoch, Alexander, Andreas Fickers, and Christian Henrich-Franke (eds.). 2013. *Airy Curtains in the European Ether: Broadcasting and the Cold War*. Baden-Baden: Nomos Verlag.

Bahktin, Mihkail M. 2002. "Forms of Time and of the Chronotope in the Novel: Notes toward a Historical Poetics." In *Narrative Dynamics. Essays on Time, Plot, Closure, and Frames*, edited by Brian Richardson, 15–24. Columbus: Ohio State University Press.

Balbi, Gabriele, and Andreas Fickers. 2020. *History of the International Telecommunication Union (ITU): Transnational Techno-Diplomacy from the Telegraph to the Internet*. Berlin: De Gruyter.

Ball, Desmond. 1991. "Improving Communications Links between Moscow and Washington." *Journal of Peace Research* 38 (2): 135–159.

Barnett, Nicholas. 2013. "'RUSSIA WINS SPACE RACE': The British Press and the Sputnik Moment, 1957." *Media History* 19 (2): 182–195.

Battaglia, Debbora. 2021. "Arresting Hospitality: The Case of the 'Handshake in Space.'" *Journal of the Royal Anthropological Institute* 18 (1): 76–89.

Beutelschmidt, Thomas. 2017. *Ost–West–Global: Das Sozialistische Fernsehen im Kalten Krieg.* Leipzig: Vistas.

Bichsel, Christine. 2020. "Introduction: Infrastructure on/off Earth." *Roadsides* 003: 1–6.

Bijker, Wiebe E., Trevor J. Pinch, and Thomas P. Hughes. 1987. *The Social Construction of Technological Systems.* Cambridge, MA: MIT Press.

Bird, Robert. 2017. "Revolutionary Synchrony: A Day of the World." *Baltic Worlds,* no. 3, 45–52.

Borowitz, Mariel. 2017. *Open Space: The Global Effort for Open Access to Environmental Satellite Data.* Cambridge, MA: MIT Press.

Bourdon, Jerôme. 2000. "Live Television Is Still Alive: On Television as an Unfulfilled Promise." *Media, Culture & Society* 22 (5): 531–556.

Bowker, Geoffrey C., and Susan Leigh Star. 1999. *Sorting Things Out: Classification and Its Consequences.* Cambridge, MA: MIT Press.

Boym, Svetlana. 2001. "Kosmos: Remembrances of the Future." In *Kosmos: A Portrait of the Russian Space Age,* edited by Adam Bartos and Svetlana Boym, 82–99. Princeton, NJ: Princeton University Press.

Bryld, Mette, and Nina Lykke. 2000. *Cosmodolphins: Feminist Cultural Studies of Technology, Animals and the Sacred.* London: Zed Books.

Buck, Susan. 2012. *The Global Commons: An Introduction.* Washington, DC: Island Press.

Burgess, Colin, and Burt Vis. 2016. *Interkosmos: The Eastern Bloc's Early Space Program.* Chichester: Springer Praxis.

Burgess, Matt. 2022. "A Mysterious Satellite Hack Has Victims Far beyond Ukraine." *Wired,* March 23, 2022. https://www.wired.com/story/viasat-internet-hack-ukraine-russia/.

Butrica, Andrew J. 1997. *Beyond the Ionosphere: Fifty Years of Satellite Communication.* Washington, DC: National Aeronautics and Space Administration (NASA).

Campbell, Robert. 1985. "Satellite Communications in the USSR." *Soviet Economy* 4 (1): 313–339.

Carey, James W. 1989. *Communication as Culture: Essays on Media and Society.* Boston: Unwin Hyman.

Caygill, Howard. 2021. "Heidegger and the Automatic Earth Image." *Philosophy Today* 65 (2): 325–338.

Chalaby, Jean K., ed. 2005. *Transnational Television Worldwide: Towards a New Media Order.* London: I. B. Taurus.

Chertok, Boris. 2009. *Rockets and People, Vol. III: Hot Days of the Cold War*. Washington, DC: National Aeronautics and Space Administration (NASA).

Child, Jack. 2008. *Miniature Messages: The Semiotics and Politics of Latin American Postage Stamps*. Durham, NC: Duke University Press.

Chu, Pey-Yi. 2020. *The Life of Permafrost: A History of Frozen Earth in Russian and Soviet Science*. Toronto: University of Toronto Press.

Cirac-Claveras, Gemma. 2022. "Re-Imagining the Space Age: Early Satellite Development from Earthly Fieldwork Practice." *Science as Culture* 31 (2), 163–186.

Clark, Evert. 1966. "Propaganda Is Called a Peril of Communications Satellites." *New York Times*, May 5, 1966, 17.

Clark, Katerina. 2011. *Moscow, the Fourth Rome: Stalinism, Cosmopolitanism, and the Evolution of Soviet Culture, 1931–1941*. Cambridge, MA: Harvard University Press.

Clarke, Arthur C. 1945. "Extra-Terrestrial Relays: Can Rocket Stations Give World-Wide Radio Coverage?" *Wireless World* 51 (10): 305–308.

Clarke, Arthur C. 1945/1998. "The Space-Station: Its Radio Applications." In *Exploring the Unknown: Selected Documents in the History of the US Civilian Space Program, Volume 3; Using Space*, edited by John M. Logsdon, Roger D. Launius, David H. Onkst, and Stephen J. Garber, 12–15. Washington, DC: National Aeronautics and Space Administration (NASA).

Clery, Daniel. 2020. "Starlink Already Threatens Optical Astronomy: Now, Radio Astronomers Are Worried." *Science*, October 9, 2020. https://www.science.org/content/article/starlink-already-threatens-optical-astronomy-now-radio-astronomers-are-worried.

Clormann, Michael, and Nina Klimburg-Witjes. 2022. "Troubled Orbits and Earthly Concerns: Space Debris as a Boundary Infrastructure." *Science, Technology, & Human Values* 47 (5), 960–985.

Cosgrove, Denis. 2001. *Apollo's Eye: A Cartographic Genealogy of the Earth in the Western Imagination*. Baltimore: Johns Hopkins University Press.

Cusack, Igor. 2005. "Tiny Transmitters of Nationalist and Colonial Ideology: The Postage Stamps of Portugal and Its Empire." *Nations and Nationalism* 11 (4): 591–612.

Damskii, Ia. 1968. "Minuta v Efire. Korotkaia Retsenziia." *Sovetskoe Radio i Televidenie*, no. 2: 10–14.

d'Arcy, Jean. 1970. "Challenge to Cooperation," *Saturday Review*, October 24, 1970, 24–25, 72–73.

Dayan, Daniel, and Elihu Katz. 1992. *Media Events: The Live Broadcasting of History*. Cambridge, MA: Harvard University Press.

Degenhardt, Wolfgang, and Elisabeth Strautz. 1999. *Auf der Suche Nach dem Europäischen Programm: Die Eurovision 1954–1970*. Baden-Baden: Nomos Verlag.

della Dora, Veronica. 2023. "From the Radio Shack to the Cosmos: Listening to Sputnik during the International Geophysical Year (1957–1958)." *Isis* 114 (1): 123–149.

Devgon, Urmila. 1970. "Intelsat: A Cooperative Partnership in Communications." *Topic: Special Issue: Life in the 21st Century*, no. 52: 23–25.

Döring, Jörg, and Tristan Thielmann. 2009. *Mediengeographie: Theorie—Analyse—Diskussion*. Bielefeld: transcript Verlag.

Downing, John. 1985. "The Intersputnik System and Soviet Television." *Soviet Studies* 37 (4): 465–483.

Downing, John D. H. 1989. "International Communications and the Second World: Developments in Communication Strategies." *European Journal of Communication* 4 (1): 99–119.

Du Maurier, George. 1878. "Edison's Telephonoscope (Transmits Light as Well as Sound), Almanac for 1879." *Punch*, December 9, 1878.

Early, L. B., L. Kumins, and J. Baer. 1966. "Business Forecasting for Communication Satellite Systems." In *Communication Satellite Systems Technology*, edited by Richard B. Marsten, 941–954. New York: Academic Press.

The Economist. 2020. "The Geopolitics of 5G: America's War on Huawei Nears Its Endgame," July 16, 2020. https://www.economist.com/briefing/2020/07/16/americas-war -on-huawei-nears-its-endgame.

Edwards, Paul N. 2006. "Meteorology as Infrastructural Globalism." *Osiris* 21 (1): 229–250.

Edwards, Paul N. 2010. *A Vast Machine: Computer Models, Climate Data, and the Politics of Global Warming*. Cambridge, MA: MIT Press.

Eko, Lyombe. 2001. "Steps toward Pan-African Exchange: Translation and Distribution of Television Programs Across Africa's Linguistic Regions." *Journal of Black Studies* 3 (31): 365–379.

Elbert, Bruce R. 2001. *The Satellite Communication Ground Segment and Earth Station Handbook*. Boston: Artech House.

Ellis, Thomas. 2019. "'Howdy Partner!': Space Brotherhood, Detente and the Symbolism of the 1975 Apollo-Soyuz Test Project." *Journal of American Studies* 53 (3): 744–769.

Emery, Walter B. 1969. *National and International Systems of Broadcasting: Their History, Operation, and Control*. East Lansing: Michigan State University Press.

Ericson, Staffan, and Kristina Riegert. 2010. *Media Houses: Architecture, Media and the Production of Centrality*. New York: Peter Lang.

Eugster, Ernest. 1983. *Television Programming across National Boundaries: The EBU and OIRT Experience*. Dedham, MA: Artech House.

Evans, Christine. 2010. "A 'Panorama of Time': The Chronotopics of Programma 'Vremia.'" *Ab Imperio* (2): 121–146.

Evans, Christine. 2016. *Between Truth and Time: A History of Soviet Central Television*. New Haven, CT: Yale University Press.

Evans, Christine, and Lars Lundgren. 2016. "Geographies of Liveness: Time, Space, and Satellite Networks as Infrastructures of Live Television in the Our World Broadcast." *International Journal of Communication* 10: 5362–5380.

Fainberg, Dina. 2021. *Cold War Correspondents: Soviet and American Reporters on the Ideological Frontlines*. Baltimore: Johns Hopkins University Press.

Farry, James, and David A Kirby. 2012. "The Universe Will Be Televised: Space, Science, Satellites and British Television Production, 1946–1969." *History and Technology* 28 (3): 311–333.

Feuer, Jane. 1983. "The Concept of Live Television: Ontology as Ideology." In *Regarding Television: Critical Approaches—An Anthology*, edited by E. Ann Kaplan, 12–22. Los Angeles: American Film Institute.

Fickers, Andreas. 2010. "The Techno-Politics of Colour Britain and the European Struggle for a Colour Television Standard." *Journal of British Cinema and Television* 7 (1): 95–114.

Fickers, Andreas, and Andy O'Dwyer. 2012. "Reading between The Lines: A Transnational History of the Franco-British 'Etente Cordiale' in Post-War Television." *VIEW Journal of European Television History and Culture* 1 (2): 1–15.

Fokin, Yuri. 1968. "Kak vse eto nachinalos [How it All Began]." *Sovietskoe radio I televidenie [Soviet Radio and Television]* 2 (February 1968): 10.

Foust, James C. 2011. "The 'Atomic Bomb' of Broadcasting: Westinghouse's 'Stratovision' Experiment, 1944–1949." *Journal of Broadcasting & Electronic Media* 55 (4): 510–525.

Fuller, R. Buckminster. 1969. *Operation Manual for Spaceship Earth*. Carbondale: Southern Illinois University Press.

Gabrys, Jennifer. 2016. *Program Earth: Environmental Sensing Technology and the Making of a Computational Planet*. Minneapolis: University of Minnesota Press.

Galison, Peter, and Elisabeth Kessler. 2019. "To See the Unseeable: Peter Galison in Conversation with Elizabeth Kessler." *Aperture*, no. 237, 72–77.

Galloway, Alexander R. 2021. *Uncomputable: Play and Politics in the Long Digital Age*. London: Verso.

Geddes, William G. 1979. "The Live Via Satellite Era." In *Intelsat Memoirs*. Washington, DC: International Telecommunications Satellite Organization.

Geppert, Alexander C. T., ed. 2018. *Limiting Outer Space: Astroculture after Apollo*. London: Palgrave Macmillan.

Geppert, Alexander C. T. 2018. "The Post-Apollo Paradox: Envisioning Limits during the Planetized 1970s." In *Limiting Outer Space: Astroculture after Apollo*, edited by Alexander C. T. Geppert, 3–26. London: Palgrave Macmillan.

Gerovitch, Slava. 2014. *Voices of the Soviet Space Program: Cosmonauts, Soldiers, and Engineers Who Took the USSR into Space*. London: Palgrave Macmillan.

Gerovitch, Slava. 2015. *Soviet Space Mythologies: Public Images, Private Memories, and the Making of a Cultural Identity*. Pittsburgh: University of Pittsburgh Press.

Gorman, Alice. 2019. *Dr. Space Junk vs. the Universe. Archaeology and the Future*. Cambridge, MA: MIT Press.

Gourne, Isabelle. 2016. " De Passer les Tensions Est-Ouest pour la Conquête de l'Espace: La Coopération Franco-Soviétique au Temps de la Guerre Froide." *Cahiers SIRICE* 2 (16): 49–67.

Graham, Stephen. 2016. *Vertical: The City from Satellites to Bunkers*. London: Verso Books.

Graham, Stephen, and Simon Marvin. 2001. *Splintering Urbanism: Networked Infrastructures, Technological Mobilities and the Urban Condition*. London: Routledge.

Graham, Stephen, and Nigel Thrift. 2007. "Out of Order." *Theory, Culture & Society* 24 (3): 1–25.

Haley, Andrew Gallagher. 1963. *Space Law and Government*. New York: Appleton-Century-Crofts.

Harvey, David. 1989. *The Condition of Postmodernity: An Enquiry into the Origin of Cultural Change*. Oxford: Blackwell.

Hay, James. 2012. "The Invention of Air Space, Outer Space, and Cyberspace." In *Down to Earth: Satellite Technologies, Industries, and Cultures*, edited by Lisa Parks and James Schwoch, 19–41. New Brunswick, NJ: Rutgers University Press.

Hecht, Gabrielle. 2011. "Introduction." In *Entangled Geographies: Empire and Technopolitics in the Global Cold War*, edited by Gabrielle Hecht, 1–12. Cambridge, MA: MIT Press.

Henke, Christopher R., and Benjamin Sims. 2020. *Repairing Infrastructures: The Maintenance of Materiality and Power*. Cambridge, MA: MIT Press.

Henrich-Franke, Christian, and Regina Immel. 2013. "Making Holes in the Iron Curtain? The Television Programme Exchange across the Iron Curtain in the 1960s and 1970s." In *Airy Curtains in the European Ether: Broadcasting and the Cold War*, edited by Alexander Badenoch, Andreas Fickers, and Christian Henrich-Franke, 177–213. Baden-Baden: Nomos Verlag.

Herzog, Herta. 1946. "Radio—The First Post-War Year." *Public Opinion Quarterly* 10 (3): 297–313.

Hewson, Martin. 1999. "Did Global Governance Create Informational Globalism?" In *Approaches to Global Governance Theory*, edited by Martin Hewson and Timothy J. Sinclair, 97–113. Albany: State University of New York (SUNY).

Hills, Jill. 2007. *Telecommunications and Empire*. Urbana: University of Illinois Press.

Hilmes, Michele. 2012. *Network Nations: A Transnational History of British and American Broadcasting*. London: Routledge.

Högselius, Per. 2013. *Red Gas: Russia and the Origins of European Energy Dependence*. London: Palgrave Macmillan.

Högselius, Per. 2022. "The Hidden Integration of Central Asia: The Making of a Region through Technical Infrastructures." *Central Asian Survey* 41 (2), 223–243.

Högselius, Per, and Yao Dazhi. 2017. "The Hidden Integration of Eurasia: East-West Relations in the History of Technology." *Acta Baltica Historiae et Philosophiae Scientiarum* 5 (2): 71–99.

Holly, Susan K., ed. 1991. *Foreign Relations of the United States, 1964–1968, Volume XXXIV, Energy Diplomacy and Global Issues.* Washington, DC: Government Printing Office. https://history.state.gov/historicaldocuments/frus1964-68v34/d100.

Howell, Elisabeth. 2022. "NASA Is Concerned about SpaceX's New Generation of Starlink Satellites." *Space.com*, February 10. https://www.space.com/nasa-collision -risk-starlink.

Hu, Tung-Hui. 2015. *A Prehistory of the Cloud.* Cambridge, MA: MIT Press.

Hughes, Thomas P. 1983. *Networks of Power: Electrification in Western Society, 1880– 1930.* Baltimore: Johns Hopkins University Press.

Hultén, Olof. 1973. "The Intelsat System: Some Notes on Television Utilization of Satellite Technology." *International Communication Gazette* 19 (1): 29–37.

Huxtable, Simon. 2022. *News from Moscow: Soviet Journalism and the Limits of Postwar Reform.* Oxford: Oxford University Press.

Ilmonen, Kari. 1996. "The Basis of It All–Technology." In *Yleisradio 1926–1996: A History of Broadcasting in Finland,* edited by Rauno Endén, 229–266. Helsinki: Finnish Historical Society.

Imre, Anikó. 2016. *TV Socialism.* Durham, NC: Duke University Press.

Innis, Harold A. 1951. *The Bias of Communication.* Toronto: University of Toronto Press.

Institute of Electrical and Electronics Engineers (IEEE). 1970. "Conference Publica- tion No. 72, Earth Station Technology." In *IEEE Conference on Earth Station Technol- ogy, 14–16 October 1970.* London: Institute of Electrical and Electronics Engineers (IEEE).

International Telecommunications Union (ITU). 1967. *Sixth Report by the Interna- tional Telecommunication Union on Telecommunication and the Peaceful Uses of Outer Space (1967).* Geneva: International Telecommunications Union (ITU).

Intersputnik. 2011. *Intersputnik: Chronicling a Long and Glorious Path.* Moscow: Smit and Hartman.

Jasanoff, Sheila. 2001. "Image and Imagination: The Formation of Global Environ- mental Consciousness." In *Changing the Atmosphere: Expert Knowledge and Environmen- tal Governance,* edited by Clark A. Miller and Paul N. Edwards, 309–337. Cambridge, MA: MIT Press.

Jasanoff, Sheila. 2004. "Heaven and Earth: The Politics of Environmental Images." In *Earthly Politics: Local and Global in Environmental Governance,* edited by Marybeth Long and Sheila Jasanoff, 31–52. Cambridge, MA: MIT Press.

Jenks, Andrew. 2018. "Transnational Utopias, Space Exploration and the Association of Space Explorers, 1972–85." In *Limiting Outer Space: Astroculture after Apollo,* edited by Alexander C. T. Geppert, 209–235. London: Palgrave McMillan.

Jenks, Andrew. 2020. "Securitization and Secrecy in the Late Cold War: The View from Space." *Kritika: Explorations in Russian and Eurasian History* 21 (3): 659–689.

Jenks, Andrew. 2021. *Collaboration in Space and the Search for Peace on Earth,* Anthem Series on Russian, East European, and Eurasian Studies. New York: Anthem Press.

Jenks, Andrew. 2021. "U.S.-Soviet Handshakes in Space and the Cold War Imaginary." *Journal of Cold War Studies* 23 (2): 100–132.

Johnson, Leland L. 1963. "The Commercial Uses of Communication Satellites." *California Management Review*, 5(3), 55–66.

Johnson, Lyndon B. 1967. "Special Message to the Congress on Communication Policy," August 14, 1967, Gerhard Peters and John T. Wooley, eds, *The American Presidency Project.* Santa Barbara: University of California, Santa Barbara, https://www.presidency.ucsb.edu/documents/special-message-the-congress-communications-policy. Accessed March 24, 2023.

Jones-Imhotep, Edward. 2017. *The Unreliable Nation: Hostile Nature and Technological Failure in the Cold War.* Cambridge, MA: MIT Press.

Jue, Melody. 2020. *Wild Blue Media: Thinking through Seawater.* Durham, NC: Duke University Press.

Jue, Melody, and Rafico Ruiz, eds. 2021. *Saturation: An Elemental Politics.* Durham, NC: Duke University.

Kildow, Judith Tegger. 1973. *INTELSAT: Policy-Maker's Dilemma.* Lexington, MA, Lexington Books.

Kilgore, Douglas De Witt. 2003. *Astrofuturism: Science, Race, and Visions of Utopia in Space.* Philadelphia: University of Pennsylvania Press.

Kurgan, Laura. 2013. *Close up at a Distance: Mapping, Technology, and Politics.* New York: Zone Books.

Larkin, Brian. 2008. *Signal and Noise: Media, Infrastructure, and Urban Culture in Nigeria.* Durham, NC: Duke University Press.

Larkin, Brian. 2013. "The Politics and Poetics of Infrastructure." *Annual Review of Anthropology* 42 (1): 327–343.

Lazier, Benjamin. 2011. "Earthrise; or, the Globalization of the World Picture." *American Historical Review* 116 (3): 602–630.

Lee, Edward G. 1970. "UNESCO Meeting on Space Communications: Legal Issues." *University of Toronto Law Journal* 20 (3): 375–379.

Leeuw, Sonja de. 2010. "Transnationality in Dutch (Pre) Television." *Media History* 16 (1): 13–29.

Lefebvre, Henri. 1974. *The Production of Space.* London: Blackwell.

Lemberg, Diana. 2019. *Barriers Down: How American Power and Free-Flow Policies Shaped Global Media.* New York: Columbia University Press.

Lerman, Rachel, and Cat Zakrzewski. 2022. "Elon Musk's Starlink Is Keeping Ukrainians Online When Traditional Internet Fails." *Washington Post*, March 19. https://www.washingtonpost.com/technology/2022/03/19/elon-musk-ukraine-starlink/.

Light, Jennifer S. 2006. "Facsimile: A Forgotten 'New Medium' from the 20th Century." *New Media & Society* 8 (3): 355–378.

Lima, Cristiano. 2022. "U.S. Quietly Paying Millions to Send Starlink Terminals to Ukraine, Contrary to SpaceX Claims." *Washington Post*, April 8. https://www

.washingtonpost.com/politics/2022/04/08/us-quietly-paying-millions-send-starlink
-terminals-ukraine-contrary-spacexs-claims/.

Logsdon, John M., Roger D. Launius, David H. Onkst, and Stephen J. Garber. 1998. *Exploring the Unknown: Selected Documents in the History of the US Civilian Space Program. Volume 3; Using Space.* Washington, DC: NASA.

Lommers, Suzanne. 2012. *Europe—On Air: Interwar Projects for Radio Broadcasting.* Amsterdam: Amsterdam University Press.

Lovejoy, Alice, and Mari Pajala, eds. 2022. *Remapping Cold War Media: Institutions, Infrastructures, Translations.* Bloomington: Indiana University Press.

Lovell, Stephen. 2015. *Russia in the Microphone Age.* Oxford: Oxford University Press.

Lundgren, Lars. 2012. "Live from Moscow: The Celebration of Yuri Gagarin and Transnational Television in Europe." *VIEW Journal of European Television History and Culture* 1 (2): 45–55.

Lundgren, Lars. 2015. "Transnational Television in Europe: Cold War Competition and Cooperation." In *Beyond the Divide: Entangled Histories of Cold War Europe,* edited by Simo Mikkonen and Pia Koivunen, 237–256. New York: Berghahn Books.

Lundgren, Lars. 2017. "(Un)Familiar Spaces of Television Production: The BBC's Visit to the Soviet Union in 1956." *Historical Journal of Film, Radio and Television* 37 (2): 315–332.

Macauley, David. 1992. "Out of Place and Outer Space: Hannah Arendt on Earth Alienation: An Historical and Critical Perspective." *Capitalism Nature Socialism* 3 (4): 19–45.

Maher, Neil M. 2017. *Apollo in the Age of Aquarius.* Cambridge, MA: Harvard University Press.

Mark, James, Artemy M. Kalinovsky, and Steffi Marung, eds. 2020. *Alternative Globalizations: Eastern Europe and the Postcolonial World.* Bloomington: Indiana University Press.

Marquardt, Alex. 2022. "Musk's SpaceX Says It Can No Longer Pay for Critical Satellite Services in Ukraine, Asks Pentagon to Pick up the Tab," *CNN,* October 14. https://www.cnn.com/2022/10/13/politics/elon-musk-spacex-starlink-ukraine

Martello, Marybeth Long, and Sheila Jasanoff. 2004. "Introduction: Globalization and Environmental Governance." In *Earthly Politics: Local and Global in Environmental Governance,* edited by Marybeth Long Martello and Sheila Jasanoff, 1–29. Cambridge, MA: MIT Press.

Massey, Doreen. 1999. "Space-Time, 'Science' and the Relationship between Physical Geography and Human Geography." *Transactions of the Institute of British Geographers* 24: 261–276.

McCormack, Derek P. 2017. "Elemental Infrastructures for Atmospheric Media: On Stratospheric Variations, Value and the Commons." *Environment and Planning D: Society and Space* 35 (3): 418–437.

McCray, W. Patrick. 2008. *Keep Watching the Skies! The Story of Operation Moonwatch and the Dawn of the Space Age.* Princeton, NJ: Princeton University Press.

McDanie, Drew, and Lewis A. Day. 1974. "INTELSAT and Communist Nations' Policy on Communications Satellites." *Journal of Broadcasting* 18 (3): 311–322.

McLuhan, Marshall. 1974. "At the Moment of Sputnik the Planet Became a Global Theater in Which There Are No Spectators but Only Actors." *Journal of Communication* 24 (1): 48–58.

McWhinney, Edward. 1977. "Review: Communication via Satellite: A Vision in Retrospect [Smith]." *American Journal of International Law* 71 (4): 834–835.

Medina, Eden. 2014. *Cybernetic Revolutionaries: Technology and Politics in Allende's Chile*. Cambridge, MA: MIT Press.

Mesiatsev, Nikolai. 2005. *Gorizonty i vertikaly moei zhizni*. Moscow: Vagrius Press.

Messeri, Lisa. 2016. *Placing Outer Space: An Earthly Ethnography of Other Worlds*. Durham, NC: Duke University Press.

Messeri, Lisa. 2020. "The Moon's Earth." *New Geographies 11, Extraterrestrial*, edited by Jeffrey S Nesbit and Guy Trangos, 77–83.

Mickiewicz, Ellen Propper. 1988. *Split Signals: Television and Politics in the Soviet Union*. Oxford: Oxford University Press.

Mihelj, Sabina, and Simon Huxtable. 2015. "The Challenge of Flow: State Socialist Television between Revolutionary Time and Everyday Life." *Media, Culture & Society* 38 (3): 332–348.

Mihelj, Sabina, and Simon Huxtable. 2018. *From Media Systems to Media Cultures: Understanding Socialist Television*. Cambridge: Cambridge University Press.

Mikkonen, Simo, and Pia Koivunen, eds. 2015. *Beyond the Divide: Entangled Histories of Cold War Europe*. New York: Berghahn Books.

Misa, Thomas J., and Johan Schot. 2005. "Introduction: Inventing Europe: Technology and the Hidden Integration of Europe." *History and Technology* 21 (1): 1–19.

Morozova, Elina, and Yaroslav Vasyanin. "International Space Law and Satellite Telecommunications." *Oxford Research Encyclopedia of Planetary Science*. December 23, 2019; Accessed March 25, 2023. https://oxfordre.com/planetaryscience/view/10.1093/acrefore/9780190647926.001.0001/acrefore-9780190647926-e-75.

Moss, Richard A. 2017. *Nixon's Back Channel to Moscow: Confidential Diplomacy and Détente*. Lexington: University Press of Kentucky.

Mueller, Milton L. 2010. *Networks and States: The Global Politics of Internet Governance*. Cambridge, MA: MIT Press.

Mustata, Dana. 2013. "Geographies of Power: The Case of Foreign Broadcasting in Dictatorial Romania." In *Airy Curtains in the European Ether: Broadcasting and the Cold War*, edited by Alexander Badenoch, Andreas Fickers, and Christian Henrich-Franke, 149–174. Baden-Baden: Nomos Verlag.

Mustata, Dana. 2019. "Architecture Matters: Doing Television History at Ground Zero." *Journal of Popular Television* 7 (2): 177–199.

Nall, Joshua. 2019. *News from Mars: Mass Media and the Forging of a New Astronomy, 1860–1910*. Pittsburgh: University of Pittsburgh Press.

Nelson, Sarah. 2021. "A Dream Deferred: UNESCO, American Expertise, and the Eclipse of Radical News Development in the Early Satellite Age." *Radical History Review*, no. 141: 30–59.

Nelson, Sarah. 2021. *Networking Empire: International Organizations, American Power, and the Struggle over Global Communications in the 20th Century.* PhD dissertation. Vanderbilt University, Nashville.

Nordenstreng, Kaarle, and Tapio Varis. 1974. "Television Traffic: A One-Way Street? A Survey and Analysis of the International Flow of Television Programme Material." In *UNESCO Reports and Papers on Mass Communication* 70: 1–65. UNESCO Reports and Papers on Mass Communication.

Ogle, Vanessa. 2015. *The Global Transformation of Time, 1870–1950.* Cambridge, MA: Harvard University Press.

Oliver, Kelly. 2015. *Earth and World: Philosophy After the Apollo Missions.* New York: Columbia University Press.

O'Toole, Thomas. 1972. "Chinese May Pull the Plug into Satellite System." *Washington Post*, January 27, 1972.

Parks, Lisa. 2005. *Cultures in Orbit: Satellites and the Televisual.* Durham, NC: Duke University Press.

Parks, Lisa. 2009. "Around the Antenna Tree: The Politics of Infrastructural Visibility." *Flow* 9 (8): 1–9. http://flowtv.org/2009/03/aroundthe-antenna-tree-the-politics-ofinfrastructural-visibilitylisa-parks-uc-santabarbara/.

Parks, Lisa. 2012. "Satellites, Oil, and Footprints: Eutelsat, Kazat, and Post-Communist Territories in Central Asia." In *Down to Earth: Satellite Technologies, Industries, and Cultures*, edited by Lisa Parks and James Schwoch, 122–137. New Brunswick, NJ: Rutgers University Press.

Parks, Lisa. 2015. "'Stuff You Can Kick!': Toward a Theory of Media Infrastructures." In *Between Humanities and the Digital*, edited by Patrik Svensson and David Theo Goldberg, 355–373. Cambridge, MA: MIT Press.

Parks, Lisa. 2018. *Rethinking Media Coverage: Vertical Mediation and the War on Terror.* London: Routledge.

Parks, Lisa. 2020. "Global Networking and the Contrapuntal Node: The Project Mercury Earth Station in Zanzibar, 1959–64." *Zeitschrift für Medien- und Kulturforschung* 11 (1): 40–57.

Parks, Lisa, and James Schwoch. 2012. *Down to Earth: Satellite Technologies, Industries, and Cultures.* New Brunswick, NJ: Rutgers University Press.

Parks, Lisa, and Nicole Starosielski, eds. 2015. *Signal Traffic: Critical Studies of Media Infrastructures.* Urbana: University of Illinois Press.

Pelton, Joseph N. 1983. "Intelsat and the Article XIV (d) Test of Significant Economic Harm or, If Someone Is Dead Would You Ask If They Were Significantly Wounded?" Presentation at the Satellite Communications Users Conference, August 22–24, 1983, St. Louis, MO.

Peters, Benjamin. 2016. *How Not to Network a Nation: The Uneasy History of the Soviet Internet*. Cambridge, MA: MIT Press.

Peters, John Durham. 1999. *Speaking into the Air: A History of the Idea of Communication*. Chicago: University of Chicago Press.

Pool, Ithiel de Sola. 1979. "Direct Broadcast Satellites and the Integrity of National Cultures." In *National Sovereignty and International Communication: A Reader*, edited by Kaarle Nordenstreng and Herbert Schiller, 120–153. Norwood, NJ: Ablex Publishing Corporation.

Poole, Robert. 2008. *Earthrise: How Man First Saw the Earth*. New Haven, CT: Yale University Press.

Potter, Simon J. 2020. *Wireless Internationalism and Distant Listening: Britain, Propaganda, and the Invention of Global Radio, 1920–1939*. Oxford: Oxford University Press.

Punch. 1892. "Reading the Stars a La Mode," August 20, 1892, 78.

Queeney, Kathryn M. 1978. *Direct Broadcast Satellites and the United Nations*. Alphen aan den Rijn: Sijthoff and Noordhoff.

Raento, Pauliina. 2011. "Introducing Popular Icons of Political Identity." *Geographical Review* 101 (1): iii–vi.

Rand, Lisa Ruth. 2018. "Falling Cosmos: Nuclear Reentry and the Environmental History of Earth Orbit." *Environmental History* 24 (1): 78–103.

Redfield, Peter. 2000. *Space in the Tropics: From Convicts to Rockets in French Guiana*. Berkeley: University of California Press.

Reid, Donald M. 1984. "The Symbolism of Postage Stamps: A Source for the Historian." *Journal of Contemporary History* 19 (2): 223–249.

Reynolds, David. 2009. *Summits: Six Meetings That Shaped the 20th Century*. New York: Basic Books.

Riccitiello, Robina. 1992. "Intersputnik Links East to West: Organization Builds its Customer Base by Offering Low Prices," *Space News*, October 5–11, 1992.

Rindzevicute, Egle. 2016. *The Power of Systems: How Policy Sciences Opened up the Cold War World*. Ithaca, NY: Cornell University Press.

Romijn, Peter, Giles Scott-Smith, and Joes Segal. 2012. *Divided Dreamworlds? The Cultural Cold War in East and West*. Amsterdam: Amsterdam University Press.

Roth-Ey, Kristin. 2011. *Moscow Prime Time: How the Soviet Union Built the Media Empire That Lost the Cultural Cold War*. Ithaca, NY: Cornell University Press.

Russ, Daniela. 2022. "'Socialism Is Not Just Built for a Hundred Years': Renewable Energy and Planetary Thought in the Early Soviet Union (1917–1945)." *Contemporary European History* 31 (4): 491–508.

Russill, Chris. 2013. "Guest Editorial: Earth-Observing Media." *Canadian Journal of Communication* 38 (3): 277–284.

Sanchez-Sibony, Oscar. 2014. "Capitalism's Fellow Traveller: The Soviet Union, Bretton Woods, and the Cold War, 1944–1958." *Comparative Studies in Society and History* 56 (2): 290–319.

Sanchez-Sibony, Oscar. 2014. *Red Globalization: The Political Economy of the Soviet Cold War from Stalin to Khrushchev.* Cambridge, MA: Cambridge University Press.

Sarkisova, Oksana. 2017. *Screening Soviet Nationalities: Kulturfilms from the Far North to Central Asia.* London: I. B Tauris.

Scannell, Paddy. 2014. *Television and the Meaning of "Live."* Cambridge, UK: Polity.

Schick, F. B. 1963. "Space Law and Communication Satellites." *The Western Political Quarterly* 16 (1): 14–33.

Schiller, Herbert I. 1969. *Mass Communications and American Empire.* Boston, MA: Beacon Press.

Schwoch, James. 2002. "Crypto-Convergence, Media, and the Cold War: The Early Globalization of Television Networks in the 1950s." Paper presented at the Media in Transition Conference, Massachusetts Institute of Technology, Cambridge, MA, May 2002. https://cmsw.mit.edu/mit2/Abstracts/MITSchwochTV.pdf Accessed April 4, 2023.

Schwoch, James. 2009. "The Curious Life of Telstar: Satellite Geographies from 10 July 1962 to 21 February 1963." In *Mediengeographie: Theorie—Analyse—Diskussion*, edited by Jörg Döring and Tristan Thielmann, 333–358. Bielefeld: transcript Verlag.

Schwoch, James. 2009. *Global TV: New Media and the Cold War, 1946–69.* Urbana: University of Illinois Press.

Schwoch, James. 2013. "'Removing Some Sense of Romantic Aura of Distance and Throwing Merciless Light on the Weaknesses of the American Life': Transatlantic Tensions of Telstar, 1961–1963." In *Airy Curtains in the European Ether: Broadcasting and the Cold War*, edited by Alexander Badenoch, Andreas Fickers, and Christian Henrich-Franke, 271–294. Baden-Baden: Nomos Verlag.

Schwoch, James. 2018. *Wired into Nature: The Telegraph and the North American Frontier.* Urbana: University of Illinois Press.

Shaw, Tony, and Denise J. Youngblood. 2014. *Cinematic Cold War: The American and Soviet Struggle for Hearts and Minds.* Lawrence: University Press of Kansas.

Shiga, John. 2013. "Sonar: Empire, Media, and the Politics of Underwater Sound." *Canadian Journal of Communication* 38 (3): 357–377.

Shumylovych, Bohdan, and Olha Povoroznyk. 2020. "Future from the Past: Imaginations on the Margins," *Ars Electronica*, https://ars.electronica.art/keplersgardens/de/imaginations/. Accessed March 27, 2023.

Siddiqi, Asif A. 2011. "Cosmic Contradictions: Popular Enthusiasm and Secrecy in the Soviet Space Program." In *Into the Cosmos: Space Exploration and Soviet Culture*, edited by James T. Andrews and Asif A. Siddiqi, 47–76. Pittsburgh: University of Pittsburgh.

Siddiqi, Asif. 2016. "Another Space: Global Science and the Cosmic Detritus of the Cold War." In *Space Race Archaeologies: Photographs, Biographies, and Design*, edited by Pedro Ignacio Alonso, 21–38. Berlin: DOM Publishers.

Siddiqi, Asif. 2019. "Dispersed Sites: San Marco and the Launch from Kenya." In *How Knowledge Moves: Writing the Transnational History of Science and Technology*, edited by John Krige, 175–200. Chicago: University of Chicago Press.

Siddiqi, Asif. 2021. "Shaping the World: Soviet-African Technologies from the Sahel to the Cosmos." *Comparative Studies of South Asia, Africa and the Middle East* 41 (1): 41–55.

Siefert, Marsha. 2011. "'Chingis Khan with the Telegraph': Communication in the Russian and Ottoman Empires." In *Comparing Empires: Encounters and Transfers in the Long Nineteenth Century*, edited by Jörn Leonhard and Ulrike von Hirschhause, 78–108. Göttingen: Vandenhoeck & Ruprecht.

Slotten, Hugh R. 2002. "Satellite Communications, Globalization and the Cold War." *Technology and Culture* 43 (2): 315–350.

Slotten, Hugh R. 2013. "The International Telecommunications Union, Space Radio Communication, and U.S. Cold War Diplomacy, 1957–1663." *Diplomatic History* 37 (2): 313–371.

Slotten, Hugh R. 2022. *Beyond Sputnik and the Space Race: The Origins of Global Satellite Communications*. Baltimore: Johns Hopkins University Press.

Smith, Delbert D. 1976. *Communication via Satellite*. Leiden: A. W. Sijthoff.

Snow, Marcellus S. 1980. "INTELSAT: An International Example." *Journal of Communication* 30 (2): 147–156.

Sorokin, Gennadii. 1968. "Oruzhie Ne Dlia Kustarei, 'Razbor Praktiki' Rubric." *Zhurnalist*, no. 1: 31.

Spier, Fred. 2019. "On the Social Impact of the Apollo 8 Earthrise Photo, or the Lack of It?" *Journal of Big History* 3 (3): 157–189.

Stanek, Lukasz. 2020. *Architecture in Global Socialism: Eastern Europe, West Africa, and the Middle East in the Cold War*. Princeton, NJ: Princeton University Press.

Star, Susan Leigh. 1999. "The Ethnography of Infrastructure." *American Behavioral Scientist* 43 (3): 377–391.

Star, Susan Leigh, and Karen Ruhleder. 1996. "Steps Toward an Ecology of Infrastructure: Design and Access for Large Information Spaces." *Information Systems Research* 7 (1): 111–134.

Starosielski, Nicole. 2015. *The Undersea Network*. Durham, NC: Duke University Press.

Tatarchenko, Ksenia. 2013. *"A House with the Window to the West": The Akademgorosk Computer Center (1958–1993)*. PhD diss., Princeton University, Princeton, NJ.

Tucker, Elizabeth. 1984. "CNN, Soviets Negotiating Television Pact." *Washington Post*, August 22. https://www.washingtonpost.com/archive/politics/1984/08/22/cnn-soviets-negotiating-tv-program-pact/741b0605-1853-4e2a-9c2d-450bbde6b6f8/. Accessed April 2, 2023.

Uricchio, William. 2004. "Storage, Simultaneity, and the Media Technologies of Modernity." In *Allegories of Communication: Intermedial Concerns from Cinema to the Digital*, edited by John Fullerton and Jan Olsson, 123–138. Bloomington: Indiana University Press.

Uricchio, William. 2008. "Television's First Seventy-Five Years: The Interpretative Flexibility of a Medium in Transition." In *Oxford Handbook of Film and Media Studies*, edited by Robert Kolker, 286–305. Oxford: Oxford University Press.

Varbansky, A. 1974. "The Development of the Broadcast Television Network in the USSR: A Translation." *Journal of the SMPTE* 83 (11): 897–900.

Varis, Tapio. 1974. "Global Traffic in Television." *Journal of Communication* 24 (1): 102–109.

Varis, Tapio. 1984. "The International Flow of Television Programs." *Journal of Communication* 34 (1): 143–152.

Varis, Tapio. 1986. "Trends in International Television Flow." *International Political Science Review / Revue Internationale de Science Politique* 7 (3): 235–249.

Vehlken, Sebastian, Christina Vagt, and Wolf Kittler. 2021. "Introduction: Modeling the Pacific Ocean." *Media+Environment* 3 (2): 1–16.

Velkova, Julia. 2020. "The Art of Guarding the Russian Cloud: Infrastructural Labour in a Yandex Data Centre in Finland." *Studies in Russian, Eurasian and Central European New Media*, no. 20: 47–63.

Vertesi, Janet. 2015. *Seeing Like a Rover: How Robots, Teams, and Images Craft Knowledge of Mars.* Chicago: University of Chicago Press.

Voelsen, Daniel. 2021. *Internet from Space: How New Satellite Connections Could Affect Global Internet Governance.* (SWP Research Paper, 3/2021). Berlin: Stiftung Wissenschaft und Politik -SWP- Deutsches Institut für Internationale Politik und Sicherheit.

Vogl, Joseph. 2008. "Becoming-Media: Galileo's Telescope." *Grey Room*, no. 29: 14–25.

Volf, Darina. 2021. "Evolution of the Apollo-Soyuz Test Project: The Effects of the 'Third' on the Interplay Between Cooperation and Competition." *Minerva* no. 59: 399–418.

Volkmer, Ingrid. 2008. "Satellite Cultures in Europe: Between National Spheres and a Globalized Space." *Global Media and Communication* 4 (3): 231–244.

Wall, Mike. 2021. "China's Tianhe Space Station Module Dodged SpaceX Starlink Satellites Twice This Year." *Space.com*, December 30. https://www.space.com/china -tianhe-space-station-maneuvers-spacex-starlink.

Washburn, Abbott. 1984. "Separate International Satellite Systems Raise Profound U.S. Policy Question." *Telematics and Informatics* 1 (4): 447–450.

Westad, Odd Arne. 2017. *The Cold War: A World History.* New York: Basic Books.

Whalen, David J. 2002. *The Origins of Satellite Communications, 1945–1965.* Washington, DC: Smithsonian Institution Press.

Whalen, David J. 2014. *The Rise and Fall of COMSAT, Technology, Business, and Government in Satellite Communications.* Basingstoke: Palgrave Macmillan.

Wickberg, Adam, and Johan Gärdebo, eds. 2022. *Environing Media.* London: Routledge.

Winston, Brian. 1998. *Media, Technology and Society: A History: From the Telegraph to the Internet.* London: Routledge.

Wolfe, Audra. 2018. *Freedom's Laboratory: The Cold War Struggle for the Soul of Science.* Baltimore: Johns Hopkins University Press.

Wolfe, Thomas. 2005. *Governing Soviet Journalism: The Press and the Socialist Person after Stalin.* Bloomington: Indiana University Press.

Wormbs, Nina. 2011. "Technology-Dependent Commons: The Example of Frequency Spectrum for Broadcasting in Europe in the 1920s." *International Journal of the Commons* 5 (1): 92–109.

Yardley, Christopher B. 2015. *The Representation of Science and Scientists on Postage Stamps: A Science Communication Study.* Canberra: Australian National University Press.

Zindel, Hannah. 2020. "Ballooning: Aeronautical Techniques from Montgolfier to Google." In *Cultural Techniques: Assembling Spaces, Texts and Collectives,* edited by Jörg Dünne, Kathrin Fehringer, Kristina Kuhn, and Wolfgang Struck, 107–127. Berlin: De Gruyter.

Zubok, Vladislav. 2009. *Zhivago's Children: The Last Russian Intelligentsia.* Cambridge, MA: Harvard University Press.

ILLUSTRATIONS

Figure 0.1 Vladimir Nesterov, *Zemlia slushaet* [The Earth Is Listening], 1965. Reproduced with permission, Valentina Nesterova.

Figure 1.1 Towers in the Sky, Bell Telephone System, "Project Telstar," n.d. Reproduced with permission, Nokia Corporation and AT&T Archives.

Figure 1.2 Ground stations during Telstar experiments, Bell Telephone System, "Project Telstar," n.d. Reproduced with permission, Nokia Corporation and AT&T Archives.

Figure 1.3 "Map of the Location of Orbita Stations" and "Diagram of Molnia-1's Coverage of the Earth," *Pravda*, October 29, 1967, 3.

Figure 2.1 Organization of "Our World" control and switching centers. Reproduced with permission, BBC Written Archives Centre.

Figure 2.2 "Our World" network map. Reproduced with permission, BBC Written Archives Centre.

Figure 4.1 Ladislav Sutnar, rendering of a planned Earth station in Nova Scotia for RCA. Reproduced with permission, Radoslav Sutnar.

Figure 4.2 Coverage map, Intelsat II, 1967. COMSAT, "Andover Earth Station," ca. 1967 (Washington, DC: COMSAT Information Office), 2.

Figure 4.3 COMSAT, "The Global Communications Satellite System," February 1971.

Figure 4.4 The Indonesian Satellite Corporation's rendition of its prospective Earth station.

Figure 4.5 Molniya I illustration in *Tekhnika Molodezhi*, July 1965.

Figure 4.6 Illustration of a Molniya satellite and Orbita Earth station by A. Minenkov.

Figure 4.7 The Intersputnik Earth station in Psary, Poland. Reproduced with permission, Intersputnik IOSC.

INDEX

Note: Page numbers followed by *f* indicate figures.

5G networks, 154
Fokin, Yuri, 46–47, 54, 61, 67
France, 16, 25, 35, 36, 39, 40, 56f,
    74–75, 79, 86–87, 93, 136, 154
Franco-Soviet cooperation, 39–40,
    74–75, 187n22
French Ariane program, 193n78
French Guiana, 10, 176n52
Friendship Games, 141, 152

Gabon, 113, 113f
Gagarin, Yuri, 19–20, 28, 41
Gagarin broadcast, 19–20, 27, 67
Galloway, Alexander R., 165n9
Geneva Plan, 25
Geodesy, 11
Geographies of liveness, 49–50, 68
Geostationary orbit, 32, 33
Geosynchronous orbit, 33. See also
    Geostationary orbit
Geppert, Alexander C. T., 10
Gerovitch, Slava, 175–176n51
Global
    integration, 71, 82
    liveness (see also liveness), 46, 60, 61
    media (infrastructure), 7, 8–9, 63,
        129, 146, 149, 154, 157
    presence, 16, 46, 48–49, 61, 67–68,
        92, 145, 179n2, 181n15
    public, 66, 148
    single global system, 9, 12, 22, 34, 55,
        59, 70, 74–76, 82, 85–86, 90, 110,
        145, 150, 154, 185n6
    televisual livness, 46
Globalization, 10–12, 40, 48, 90, 93,
    108, 121–123, 127, 143–145, 147,
    149, 154
    and space, 10–11, 70, 90, 112, 117,
    143
    Soviet relationship to globalization,
        108, 186n11
Global South
    access and equity, 157

building housing and other landmark
    buildings (1970s), 106
decolonization and activism, 135
Earth stations, 195n28
insecure alliances and changing
    loyalties, 109
Intelsat promotional materials, 118
opposition to US economic and
    technological hegemony, 74
and the Soviet Union, 43
"Golden spike," railway metaphor 58
Goldstein, Irving, 151
Goonhilly Downs, 34, 35, 35f, 36,
    177n62
Gorbachev, Mikhail, 108, 141
Gorizont satellites, 131, 136
Gorky, Maxim, 63
Gosteleradio, 28, 41
Graham, Stephen, 72, 186n12, 186n14
Great Patriotic War, 39
Germany, 25
    East, 56f, 78, 170n58
    West, 56f, 86
Greece, 105, 113f, 114
Ground stations, 159, 193n3. See also
    Earth stations
"Ground Stations for Space
    Telecommunications" (RCA
    pamphlet), 93–96
GT&E International Systems, 82

Harris Corporation, 132
Hecht, Gabrielle, 13
Heidegger, Martin, 163n2
Hidden integration, 12, 140, 169n51,
    185n9
High-Earth orbit, 31, 32
Hilmes, Michele, 30
Högselius, Per, 128
Horizontality of space media, 157
Hotline (new DCL line between
    Washington and Moscow),
    130–131, 140, 146–147

Infrastructures Series

Edited by Paul N. Edwards and Janet Vertesi